WASTE MANAGEMENT SERIES 6

BIOGRANULATION TECHNOLOGIES FOR WASTEWATER TREATMENT

Waste Management Series

Volume 1: Waste Materials in Construction: Science and Engineering of Recycling for Environmental Protection
G.R. Woolley, J.J.J.M. Goumans and P.J. Wainwright (Editors)

Volume 2: Geological Disposal of Radioactive Wastes and Natural Analogues
W. Miller, R. Alexander, N. Chapman, I. McKinley and J. Smellie

Volume 3: Principles and Standards for the Disposal of Long-lived Radioactive Wastes
N. Chapman, S. McCombie (Editors)

Volume 4: Solid Waste: Assessment, Monitoring and Remediation
I. Twardowska, H.E. Allen, A.F. Kettrup, W.J. Lacy

Volume 5: Olive Processing Waste Management: *Literature Review and Patent Survey, Second Edition*
Michael Niaounakis and Constantinos P. Halvadakis

WASTE MANAGEMENT SERIES 6

BIOGRANULATION TECHNOLOGIES FOR WASTEWATER TREATMENT

Joo-Hwa Tay, PhD, PE
Stephen Tiong-Lee Tay,[†] PhD
Yu Liu, PhD
Kuan-Yeow Show, PhD
Volodymyr Ivanov, PhD

School of Civil and Environmental Engineering, Nanyang Technological University, Singapore

ELSEVIER

Amsterdam – Boston – Heidelberg – London – New York – Oxford
Paris – San Diego – San Francisco – Singapore – Sydney – Tokyo

Elsevier
The Boulevard, Langford Lane, Kidlington, Oxford OX5 1GB, UK
Radarweg 29, PO Box 211, 1000 AE Amsterdam, The Netherlands

First edition 2006

British Library Cataloguing in Publication Data
A catalogue record for this book is available from the British Library

Library of Congress Cataloging-in-Publication Data
A catalog record for this book is available from the Library of Congress

ISBN-13: 978-0-08-045022-3
ISBN-10: 0-08-045022-9
ISSN: 1478-7482

For information on all Elsevier publications
visit our website at books.elsevier.com

Printed and bound in The United Kingdom

06 07 08 09 10 10 9 8 7 6 5 4 3 2 1

In Remembrance to Stephen Tiong-Lee Tay

We wish to dedicate this book to the major author, Stephen Tiong-Lee Tay. Stephen suddenly passed away on 29 July 2005, several days before the finishing of the book. He established research team on the granulation studies in Nanyang Technological University (NTU) and created there a spirit of cooperation and friendship. He has made significant contributions to the world's studies on granulation, especially in microbiology and biotechnology of the granules degrading toxic compounds. He postulated and proved that the aerobic granulation technology could overcome the disadvantages associated with the use of carrier materials in traditional cell immobilization systems. The microbial granulation research team and his students in NTU continue the development of Stephen's ideas. For all of us, who had the privilege of knowing, interacting, and working with Stephen, he was more than a brilliant and dedicated Professor, he was a dear friend, mentor, and coach, he has touched the lives of many of us.

Contents

Preface

Microbial self-aggregation, in which microbial cells are organized into dense and fast settling granules with a diameter from 0.5 to 10 mm, is extensively studied due to its practical importance in both anaerobic and aerobic biological wastewater treatment. Anaerobic and aerobic microbial granules have different properties and applications and are considered separately in this book.

Formation of anaerobic granules is discussed in Chapter 1. There are many theoretical explanations, which must be taken into account in practical performance of granular anaerobic wastewater treatment. It is favorable for the microorganisms to be very close to each other in the granule in order to achieve high substrate conversion rate. Possible advantages of microorganisms in anaerobic granule in comparison with flocculated or suspended microorganisms are as follows:

1. aggregation leads to heterogeneous community and facilitates syntrophic relationships, especially interspecies hydrogen and formate transfer;
2. granulation protects cells from predators, such as anaerobic ciliates;
3. under unfavorable conditions for growth (e.g. extreme pH), a more favorable micro-environment can be maintained within the aggregates so that metabolism can be sustained;
4. the diffusion of substrates and fermentation products can be facilitated due to the formation of the channels in the granule.

Most valuable data for the practice are given in Chapter 2, where the effects of such factors as temperature, pH, upflow velocity, hydraulic retention time, organic loading rate, and type of substrate on anaerobic granulation are described. The real applications of anaerobic granulation are described in Chapter 3. The reader can find the description of granulation process in upflow anaerobic sludge blanket reactor (UASB), expanded granular sludge bed reactor (EGSBR), hybrid anaerobic reactor (HAR), anaerobic continuous stirred tank reactor (ACSTR), anaerobic baffled reactor (ABR), anaerobic sequencing batch reactor (ASBR), and anaerobic migrating blanket reactor (AMBR). The main problem associated with the granular sludge systems is the long start-up period required for the development of anaerobic granules. In cases where a reactor is seeded with flocculant sludge, obtained from municipal wastewater sludge digesters, it usually takes several months or even a much longer period before the system can be operated. In order to reduce the lengthy start-up of granular sludge-based systems, technologies for enhanced and rapid production of anaerobic granules are highly desirable and sought after. Another possibility of rapid start-up is the use of granular sludge from in-operating reactors as the seeds. This has the advantage of being able to achieve the desired performance within a short start-up period. However, the availability of granular seed sludge is limited, and the costs for purchase and transportation of the seeds can be high.

A major part of this book is devoted to aerobically grown microbial granules, which can be used or are used in the wastewater treatment. Advantages of aerobic wastewater treatment using microbial granules instead of conventional flocs of activated sludge are retention of granulated biomass in a reactor, diversity of physiological functions of microorganisms in the granule, and resistance of the microorganisms inside the granule to toxic substances. Aerobic granulation is a gradual process from seed sludge to compact aggregates, further to granular sludge, and finally to mature granules. To accelerate industrial application of the aerobic granulation technology, a sound understanding of the mechanisms behind aerobic granulation is highly desirable. Mechanisms of granulation and factors affecting aerobic granulation are discussed in Chapters 4 and 5. Such aspects of microbial self-immobilization as hydrophobic interactions, role of exopolysaccharides and other exopolymers in aerobic granulation, role of hydrodynamic shear force and selection pressure, substrate composition, organic loading, feast–famine regime, feeding strategy, concentration of dissolved oxygen, reactor configuration, solids retention time, cycle time, settling time, and exchange ratio are discussed in these chapters. In sequencing batch reactor, three major factors of selection pressure had been identified: the settling time, the volume exchange ratio, and the discharge time.

Aerobic granules, which are usually spheres or ellipsoids with size from 0.2 to 7 m have complex structure including radial inclusions, concentric layers, and central core. The granules are covered with filamentous, smooth, or skin-like surface, which is dominantly hydrophobic or hydrophilic. The interior of a granule is gel-like matrix, containing black matter or gas vesicule in central part of a big dense granule. There were found layers and microaggregates of specific microorganisms connected with the channels facilitating diffusion of substrates and products of metabolism. There are a layer of anaerobic bacteria and a core of lysed biomass in the central part of aerobically grown microbial granules. These structural elements of the granules together with the principles of structural optimization are described in Chapter 6.

Microbial diversity of aerobic granules, described in Chapter 7, was studied using cloning–sequencing method, amplified ribosomal DNA restriction analysis (ARDRA), and fluorescence *in situ* hybridization (FISH) with specific oligonucleotide probes. The analysis of the microbial community residing in the aerobically grown granule can provide information on the microorganisms responsible for granule formation, maintenance, and activity. This knowledge can be used to better the control of aerobic granulation. Data on physiological diversity, first of all, on the presence of aerobic, facultative-anaerobic and anaerobic microorganisms in the granules, were derived from identification of major microbial components of the granules. The important aspects of microbiology of microbial granules are presence of pathogens, determining biosafety of the wastewater treatment, and gliding bacteria, which are probably important microorganisms for the formation and stability of the granules.

One of the main problems of environmental engineering is removal of phosphate and ammonia/nitrate from the wastewater. Aerobically grown microbial granules are able to remove nitrogen and phosphorus from the wastewater as shown in Chapter 8. The problems encountered in the suspended growth nutrient-removal system, such as sludge bulking, large treatment plant space, washout of nitrifying biomass, secondary P release in a clarifier, higher production of waste sludge, would be overcome by developing N-removing and P-accumulating granules. A more compact and efficient granule-based biotechnology would be expected for high-efficiency N and P removal.

Together with the removal of nutrients, aerobically grown microbial granules can be applied for the biodegradation of toxic organic compounds. Advantages of microbial granules in the treatment of industrial toxic wastewater, containing phenol, are discussed in Chapter 9. Structure of these granules, their microbial content, and its response to the load of phenol are discussed aiming to find optimal strategy for the treatment of toxic wastewater with microbial granules.

One potential disadvantage of aerobic granulation is the long start-up period of granule formation from the flocs of activated sludge. Another potential disadvantage is the risk of accumulation of pathogenic microorganisms in the granule because of two reasons: 1) cells are aggregated mainly due to hydrophobic interactions and there may be accumulation of strains with high cell hydrophobicity in the granule; 2) bacterial strains with high cell surface hydrophobicity are often pathogenic ones. Addition into the reactor safe microbial cultures selected for fast formation of the granules can be used to solve these problems. Chapter 10 is devoted to the selection and use of microbial seeds (inoculum) to start-up safe granulation process. Different principles can be used in selection: strong self-aggregation of cells of one species; coaggregation of cells of different species; enrichment culture of fast-settling cells, or cells with high cell surface hydrophobicity. As shown in this chapter, application of microbial seeds for granulation can reduce start-up period from 14–21 to 2–7 days.

The conventional methods for heavy metal removal from aqueous solution include precipitation with lime or other chemicals, chemical oxidation and reduction, ion-exchange, filtration, electro-chemical treatment, reverse osmosis filtration, evaporative recovery, and solvent extraction. However, when the heavy metal concentrations in the wastewater are low, these processes would have some problems of incomplete heavy metal removal, high reagent or energy consumption, generation of toxic sludge or other wastes. Aerobic granules with strong and compact microbial structure would be a novel biosorbent for metal ion removal from a liquid solution. Biosorption of soluble heavy metals by aerobic granules is described in Chapter 11.

Mechanisms of aerobic granulation are finally not known. Physiological and biological diversity of the granules must be studied in more detail to understand the formation and functions of the granules. Such importance for the practical application property as granules stability was not explained yet in terms of mathematical model and reliable prediction. Microbial inoculum of fast-aggregating cells can be used for the facilitation granulation but biosafety, activity of pure cultures, and their domination in the granules must be studied in practical applications.

The book is covering almost all aspect of formation and use of microbial granules in the wastewater treatment. The data on aerobic microbial granulation are related mostly to laboratory systems because there are just few pilot systems in the world using aerobic microbial granules and there is no one constructed industrial facility using aerobic microbial granulation yet. However, by the analogy with anaerobic granulation which is used now worldwide, it would be possible to predict wide applications of aerobic granulation. The authors hope that this book will help researchers and engineers to develop these new biotechnologies of wastewater treatment based on aerobic granulation.

Joo-Hwa Tay
Stephen Tiong-Lee Tay
Yu Liu
Kuan-Yeow Show
Volodymyr Ivanov

Contributors

Ivanov Volodymyr, *PhD*
Associate Professor, School of Civil and Environmental Engineering, Nanyang Technological University, Singapore, E-mail: cvivanov@ntu.edu.sg

Liu Yu, *PhD*
Associate Professor, School of Civil and Environmental Engineering, Nanyang Technological University, Singapore, E-mail: CYLiu@ntu.edu.sg

Show Kuan-Yeow
Associate Professor, School of Civil and Environmental Engineering, Nanyang Technological University, Singapore, E-mail: CKYSHOW@ntu.edu.sg

Tay Joo-Hwa, *PhD, PE*
Professor, School of Civil and Environmental Engineering, Nanyang Technological University, Singapore, E-mail: CJHTAY@ntu.edu.sg

Tay Stephen Tiong-Lee, *PhD*
late Associate Professor, School of Civil and Environmental Engineering, Nanyang Technological University, Singapore

Chapter 1

Mechanisms and Models for Anaerobic Granulation

Kuan-Yeow Show

Introduction

The upflow anaerobic sludge blanket (UASB) reactor is increasingly gaining popularity for high strength organic wastewater treatment because of its high biomass concentration and rich microbial diversity (Lettinga et al., 1980; Hulshoff Pol et al., 1988; Fang et al., 1995; Schmidt and Ahring, 1996; Wu et al., 2001). High biomass concentration and rich microbial diversity give rise to rapid contaminant degradation, implying that highly concentrated or large volumes of organic waste can be treated in compact UASB reactors. Comparing to other anaerobic technologies, such as anaerobic filter, anaerobic sequencing batch reactor, anaerobic expanded bed, and fluidized bed reactors, a unique feature of the UASB system is its dependence on biogranulation process. It appears that anaerobic granular sludge is a core component of a UASB reactor. The granules are generally dense and enriched with multispecies microbial communities. None of the individual species in the granular ecosystem is capable of degrading complex organic wastes separately.

One major drawback of UASB reactors is its extremely long start-up period, which generally requires between 2 and 8 months for successful development of granular sludge. To reduce the space–time requirements and leading to a cheaper treatment of high strength wastes, strategies for expediting granules development are highly desirable for UASB systems. In achieving such a purpose, a thorough understanding of the mechanisms for anaerobic granulation is essential. This chapter attempts to review the existing mechanisms and models for anaerobic granulation in UASB systems, and also tries to build up a general model for anaerobic granulation.

Physico-chemical Models

Microbial adhesion or self-immobilization is regarded as the onset of anaerobic granulation process, and can be defined in terms of the energy involved in the interaction of bacterium-to-bacterium or bacterium-to-solid surface. In a thermodynamic sense, when one bacterium approaches another, the interactions involve repulsive electrostatic force, attractive van de Waals force, and repulsive hydration interaction. Some authors analyzed the granulation mechanism in terms of energy involved in the adhesion itself, due to the physico-chemical interactions between cells walls or between cells walls and alien surfaces. Factors like hydrophobicity and electrophoretic mobility are objectively taken into account. Based on the thermodynamics, some physico-chemical models for anaerobic granulation have been developed, those include inert nuclei model, selection pressure model, multivalence positive ion-bonding model, ECP bonding model, synthetic and natural polymer-bonding model, secondary minimum adhesion model, local dehydration and hydrophobic interaction model, and surface tension model.

Inert Nuclei Model

The inert nuclei model for anaerobic granulation was initially proposed by Lettinga et al. (1980). In the presence of inert microparticles in a UASB reactor, anaerobic bacteria could attach onto the particle surfaces to form initial biofilm, namely embryonic granules. Subsequently, mature granules can be further developed from the growth of these attached bacteria under given operating conditions. The inert nuclei model suggests that the presence of nuclei or microsize biocarrier for bacterial attachment is a first step towards anaerobic granulation. The inert nuclei model was supported by experimental evidence such that addition of zeolite or hydro-anthracite particles with a diameter of 100 μm into inoculated sludge seemed to be effective in promoting the formation of anaerobic granules (Hulshoff Pol, 1989). Water absorbing polymer (WAP) particles were also used to enhance granulation (Imai, 1997). The WAP is a pulverulent resin, which swells in water and exhibits a complex network structure, which can provide more surfaces for microbial attachment and growth than other inert particles. The laboratory-scale experiments indicated that the contact between particles and biomass could be improved since the WAP has lower density than sand and other inert materials (Imai, 1997).

Selection Pressure Model

The basis of anaerobic granulation had been proposed as a continuous selection of sludge through washing out light and dispersed bioparticles and retaining heavier biomass in the reactors (Hulshoff Pol et al., 1988). The selection pressure model suggests that microbial aggregation in UASB reactor appears to be a protective microbial response against high selection pressures. In UASB reactors, selection pressure is created by upflow liquid flow pattern. It had been reported that under very weak hydraulic selection pressure operating conditions, no anaerobic granulation was observed (Alphenaar et al., 1993; O'Flaherty et al., 1997). Rapid development of anaerobic granules could be accomplished through a purely physical aggregation from the hydraulic stress applied on the anaerobic flocculant sludge (Noyola and Mereno, 1994). The results showed that flocculant anaerobic sludge could be converted into a relatively active granular sludge by enhancing agglomeration through only short hydraulic stress of less than 8 h. Arcand et al. (1994) also reported that the liquid upflow velocity had a significant positive effect on mean granule size, but the effect on specific washout rate of smaller particles was marginal. It is very likely that relatively high selection pressure in terms of upflow liquid velocity is favorable for rapid development of anaerobic granules.

Attrition Model

Attrition model proposed that granules originate from fines formed by attrition and from colonization of suspended solids from the influent

(Pereboom, 1994). According to Pereboom (1994), increase in granule size is only due to microbial growth and therefore the concentric layers observed on sliced granules are related to small fluctuations in growth conditions.

It was reported that the most significant process limiting the maximum granule size in normal operation is the regular discharge of surplus biomass. Reactor turbulence and internal gas production appeared to have no influence on the size distribution. The shear forces due to liquid and gas turbulence are not responsible for breaking or disintegrating of granules, and only cause attrition of small particles from the granules. The attrition is not expected to be significant to the removal of large granules.

According to Pereboom (1994), the granular size distribution in UASB reactors seems to be the result of growth from small particles (being washed into the reactor or developed in the reactor by attrition) into larger granules and the removal of representative amounts of granules from all size classes by sludge discharge. Besides, wastewaters of high concentrations of suspended solids would result in narrow granule size distributions, while influent of little or no suspended solids would lead to good distribution of size.

Multivalence Positive Ion-bonding Model

As bacteria have negatively charged surfaces under normal pH conditions, a basic idea to expedite anaerobic granulation is to reduce the electrostatic repulsion between negatively charged bacteria by introducing multivalence positive ion, such as calcium, ferric, aluminum, or magnesium ions into the seed sludge. It had been reported that reduced electrostatic repulsion between bacteria would promote anaerobic granulation (Mahoney et al., 1987; Schmidt and Ahring, 1993; Yu et al., 2001a). Addition of Ca^{2+} in the range of 80–200 mg/l, Mg^{2+} of 12–120 mg/l, or Al^{3+} of 300 ml/l increased the rate of anaerobic granulation in UASB reactors (Schmidt and Ahring, 1993; Teo et al., 2000; Yu et al., 2001b). However, high calcium concentration of above 500 mg/l (Guiot et al., 1988; Thiele et al., 1990) or 600 mg/l (Yu et al., 2001a) was found detrimental to anaerobic granulation. High calcium concentrations also cause serious problems, e.g. precipitation and accumulation of calcium in anaerobic granules, as well as reduced microbial activity of granules.

The multivalence positive ion-bonding model is based on a simple electrostatic interaction between negatively charged bacteria and positive ion. The DLVO[1] theory postulates that when two surfaces carry a charge of the same sign, there is a free energy barrier between them, which acts as a repulsive force. This force could seriously prevent approach of one cell to another. A positive ion added to sludge would partially neutralize the negative charges on bacterial surfaces by adsorption, causing a significant reduction in the electrical repulsion between bacteria. The positive ion hence initiates cell-to-cell interaction which is a crucial initiation towards granulation.

In addition, the multivalence positive ion could also compress the double layer to promote cell aggregation (Zita and Hermansson, 1994). Moreover, the multivalence positive ion may promote sludge granulation by bonding with extracellular polymers (ECPs), and high affinity between ECPs and calcium ion had been reported (Forster and Lewin, 1972; Rudd et al., 1984). This implies that calcium ion may bridge ECPs to ECPs and/or link cells to ECPs to form an initial three-dimension structure of microbial community, in which bacteria could grow further.

ECP Bonding Model

The ECPs can mediate both cohesion and adhesion of cells, and play a vital role in maintaining structural integrity of microbial matrix. On the other hand, the metabolic blocking of exopolysaccharides synthesis would prevent microbial aggregation (Schmidt and Ahring, 1994; Cammarota and Sant'Anna, 1998). It had been reported that ECPs could change the surface negative charge of the bacteria, and thereby bridge two neighboring cells physically to each other, and with other inert particulate matters (Shen et al., 1993; Schmidt and Ahring, 1994, 1996). Chen and Lun (1993) observed that increasing the organic loading rate resulted in significant growth of *Methanosarcina* which secreted much more ECPs to form larger clumps, and subsequently *Methanothrix* tended to fill in the *Methanosarcina* clumps.

---

[1]DLVO theory accounts for the interaction between charged colloidal particles. It is based on the sum of a van der Waals attractive potential and a screened electrostatic potential arising from the "double layer" potential screened by ions in solution.

Synthetic and Natural Polymer-bonding Model

Synthetic polymers have been widely used in water coagulation and flocculation processes, and can significantly promote particle agglomeration. Similarly, the synthetic polymers can also be applied to expedite development of anaerobic granules. It was found that the supplementation of polymer Chitosan, which has a similar structure to polysaccharides, significantly enhanced the formation of anaerobic granules in the UASB-like reactors. Granulation rate in the Chitosan-containing reactor was 2.5-fold higher than that in the control reactor without addition of the polymer, while the specific activities of methane production were comparable in both reactors (El-Mamouni et al., 1998). In fact, it is not surprising to obtain such results since freely moving polymeric chains may form a bridge between cells, and this would facilitate the formation of initial microbial nuclei, which is the initial step towards granulation.

Kalogo et al. (2001) used water extract of *Moringa oleifera* seeds (WEMOS) to enhance the start-up of a UASB reactor treating domestic wastewater, and they found that the dosage of WEMOS in the feed favored the aggregation of coccoid bacteria and growth of microbial nuclei, which are precursors of anaerobic granulation. WEMOS, as a kind of natural polymers, is known to be effective in flocculating organic matter. Adsorption of WEMOS on the surface of the dispersed bacteria and neutralization of their surface charges would be a principal mechanism to promote anaerobic granulation.

Recently, Show et al. (2004); Wang et al. (2004) investigated the influence of a coagulant polymer on start-up, sludge granulation and the associated reactor performance in laboratory-scale UASB reactors. A control reactor R1 was operated without added polymer, while the other three reactors designated R2, R3, and R4 were operated with polymer concentrations of 5 mg l^{-1}, 10 mg l^{-1}, and 20 mg l^{-1}, respectively. The experimental results indicated that adding the polymer at a concentration of 20 mg l^{-1} markedly reduced the start-up time. The time required to reach stable treatment at an organic loading rate (OLR) of 4.8 g COD l^{-1} d^{-1} was reduced by more than 36% (R4) as compared with both R1 and R3, and by 46% as compared with R2. R4 was able to handle an OLR of 16 g COD l^{-1} d^{-1} after 93 days of operation, while R1, R2, and R3 achieved the same loading rate only after 116, 116, and 109 days, respectively. Compared with the control reactor, the start-up time of R4 was shortened by about 20% at this OLR. Granule characterization indicated that the granules developed in R4 with 20 mg l^{-1} polymer exhibited the best settleability and methanogenic activity at all OLRs. The organic loading capacities of the reactors were also increased by the polymer addition. The maximum organic loading of the control reactor (R1) without added polymer was 19.2 g COD $l^{-1}d^{-1}$, while the three polymer-assisted reactors attained a marked increase in organic loading of 25.6 g COD $l^{-1}d^{-1}$.

The findings by Show et al. (2004); Wang et al. (2004) demonstrated that adding the cationic polymer could result in shortening of start-up time and enhancement of granulation, which may in turn lead to improvement in organics removal efficiency and loading capacity of the UASB system. The authors hypothesized that positively charged polymer form bridges among the negatively charged bacterial cells through electrostatic charge attraction. The bridging effect would enable greater interaction between biosolids resulting in preferential development and enhancement of biogranulation in UASB reactors.

Secondary Minimum Adhesion Model

Secondary minimum adhesion model is based on the DLVO theory for colloidal particles, which proposes that reversible adhesion takes place in the secondary minimum of the DLVO free energy curve. The Gibbs energy of the reversible adhesion is relatively small, and there is always a separation distance between the two adhering bacteria. Thus, the reversible adhesion can change to irreversible adhesion at the primary minimum by overcoming the energy barrier or by protruding fibrils or fimbriae, which bridge the gap between bacteria (Rouxhet and Mozes, 1990). The secondary minimum adhesion model accounts for both the surface charge and the surface energy or hydrophobicity, which are relevant to long- and short-range forces. It appears from this model that anaerobic granulation would start from the self-immobilization of bacteria through reversible and followed by irreversible microbial interaction. It should be realized that the secondary minimum adhesion model merely looks into the thermodynamic aspects of bacterial interaction, thus the real meaning of this model is somewhat limited with respect to a biologically defined engineering rector.

Local Dehydration and Hydrophobic Interaction Model

Under normal pH conditions, the outer surfaces of bacteria are hydrated. Such a water layer on the surfaces of bacteria would prevent one bacterium to approach another. It has been believed that under normal physiological conditions, strong hydration repulsion is the main force keeping the cells apart, thus local dehydration of the short-distance-apart surfaces would be a prerequisite for cell-to-cell aggregation. The local dehydration and hydrophobic interaction model as proposed by Wilschut and Hoekstra (1984) shows that when bacterial surfaces are strongly hydrophobic, irreversible adhesion will occur. Theoretically, increasing the hydrophobicity of cell surfaces would cause a corresponding decrease in the excess Gibbs energy of the surface, which in turn promotes cell-to-cell interaction and further serves as driving force for cell self-separation from liquid phase (van Loosdrecht et al., 1987; Rouxhet and Mozes, 1990). There is experimental evidence showing that the hydrophobicity of bacterial surface plays a crucial role in initiating anaerobic granulation (Mahoney et al., 1987; Wu et al., 1991; Tay et al., 2000a,b).

Cell hydrophobicity can be quantified by the measurement of water contact angle (Mozes and Rouxhet, 1987; van Loosdrecht et al., 1987). The hydrophobicity of microorganisms may be roughly classified into three categories with respect to the water contact angle (Mozes and Rouxhet, 1987): hydrophobic surface with a contact angle greater than 90°, medium hydrophobic surface with a contact angle in between 50 and 60°, and hydrophilic surface with a contact angle below 40°. Most acidogens exhibit hydrophilic characteristics with a water contact angle less than 45°, however acetogens and methanogens isolated from anaerobic granules show a high surface hydrophobicity with a water contact angle greater than 45° (Daffonchio et al., 1995). The local dehydration and hydrophobic interaction model provides a physico-chemical elucidation explaining why acidogens are most often situated in outer layer of anaerobic granules.

Surface Tension Model

According to the thermodynamic theory, microbial granulation is a creation process for a new granule–liquid interface by disrupting preexisting individual bacteria–liquid interface, and a molecular contact between the two adhering bacterial surfaces is involved. The free energy of adhesion

(ΔG_{adh}) can be expressed as follows (Rouxhet and Mozes, 1990):

$$\Delta G_{\text{adh}} = 2(r_{\text{c}}^{1/2} - r_{\text{l}}^{1/2})(r_{\text{l}}^{1/2} - r_{\text{s}}^{1/2})$$

where,

r_{c} is the surface free energy of bacteria,
r_{l} is the surface free energy of liquid, and
r_{s} is the surface free energy of inert particle.

This equation shows that if the surface free energy of bacteria is lower that that of the liquid, the free energy of aggregation decreases and aggregation is favored with decreasing surface energy of the inert carrier. The opposite trend would occur if the surface energy of bacteria is higher than that of the liquid. In fact, the above thermodynamic equation is a theoretical basis of the surface tension model. It was found in a UASB reactor that aggregation of hydrophilic cells was enhanced at low liquid surface tension, while the opposite was true for hydrophobic cells (Thaveesri et al., 1995). Depending on the liquid surface tension (γ) in the UASB reactor, bacteria may grow in rather loose associations, in multilayered granules ($\gamma < 50$ mN/m), or in mixed conglomerates ($\gamma > 56$ mN/m) (Thaveesri et al., 1995; Grootaerd et al., 1997).

Consideration on the Physico-chemical Models

The discussion so far seems to suggest that each physico-chemical model accounts for contribution of only one or two factors to the initial granulation process in UASB reactor. As these factors exert their influences under specific environmental conditions and in specific steps during the entire granulation process, the physico-chemical models provide only simple descriptions on anaerobic granulation.

The inert nuclei model can be easily understood with a hypothesis that the formation of UASB granules is favored by the presence of inert particles in the reactor. However, there was evidence that anaerobic granules could be developed even without adding any inert materials (Thiele et al., 1990). It should be realized that besides attachment on solid surfaces, self-immobilization of bacteria can also lead to formation of microbial aggregates.

With respect to the multivalence positive ion-bonding model, some studies had shown that calcium ion did not contribute to sludge granulation (Guiot et al., 1988) and that a high concentration of magnesium ion caused disintegration of granules (Schmidt and Ahring, 1993). A research in membrane fusion indeed indicated that Ca^{2+} might cause conformational changes of some surface proteins or polypeptide groups that could interact with two surfaces and bridge them together (Papahadjopoulos et al., 1990). On the other hand, it had been proposed that the beneficial effect of calcium addition on anaerobic granulation was probably due to the calcium-induced dehydration and fusion of bacterial surfaces (Teo et al., 2000). The calcium-induced cell fusion might initiate the formation of cell cluster, which acts as microbial nuclei of anaerobic granulation.

In the secondary minimum adhesion model, the DLVO theory is unable to make predictions at short distances due to breakdown of the computation of electrical interactions. It also neglects the forces which are important at short distances, such as hydrogen bonding and other effects involved in solution and hydrophobic bonding (Rouxhet and Mozes, 1990). While in the local dehydration and surface tension models, bacterial granulation is oversimplified to a purely thermodynamic process. Such a simple description is usually inadequate, as microbial aggregation is a very complex biological phenomenon and many unidentified factors are believed to be involved. It seems impossible to develop a pure thermodynamic model with satisfactory confidence level. The fact that bacteria cannot be simply treated as physically defined dead colloidal particles, and bacteria indeed have no well-defined surface boundary, simple geometry, or uniform molecular surface composition, the physico-chemical forces alone are not able to completely explain the entire microbial granulation process. It is thus suggested that the physico-chemical phenomena involved in microbial granulation ought to be related to the biological triggers controlling the granulation.

Structural Models

Anaerobic granulation is a complex process, in which biological factors are involved other than physico-chemical forces. In the past two decades significant research progress had been made in understanding microbiological characteristics of UASB granules and interactions among different microbial species in the granules. In view of the development, a series of structural models for anaerobic granulation has been developed to interpret the observed phenomena.

Capetown Model

Like the polymer-bonding model as discussed earlier, the Capetown model suggests that ECPs are produced by *Methanobacterium* strain AZ, a hydrogen-utilizing methanogen (Palns et al., 1987; Sam-Soon et al., 1988). Under the conditions of high hydrogen partial pressure and limited cysteine, the amino acids (except cysteine) would be over-secreted. Excessive amino acids could induce ECPs formation, and consequently *Methanobacterium* strain AZ and other genera are enmeshed in the ECPs matrix, which in turn lead to the initiation of anaerobic granulation. In the Capetown model, the overproduction of ECPs is considered a key initiation of anaerobic granulation.

Spaghetti Model

Based on microstructure of UASB granules observed under scanning electron microscope, Wiegant (1998) proposed a spaghetti model for anaerobic granulation. This model hypothesizes that development of UASB granules is initiated by attachment of filamentous *Methanosaeta* on precursors, followed by a formation of a three-dimensional network through a branched-growth process. Other bacteria, such as *Methanosarcina*, could be easily entrapped in this network (Sanchez et al., 1994; Wu et al., 1996). The structured aggregates further develop through cellular multiplication of the entrapped bacteria, and become denser and spherical by the action of hydrodynamic shear force attributed to upflow liquid and biogas. It must be emphasized that in the spaghetti model, formation of the structured aggregate is a crucial stage of the overall granulation process.

Syntrophic Microcolony Model

The bioconversion of organics into methane proceeds through a series of complex biochemical changes, and little is known about the individual

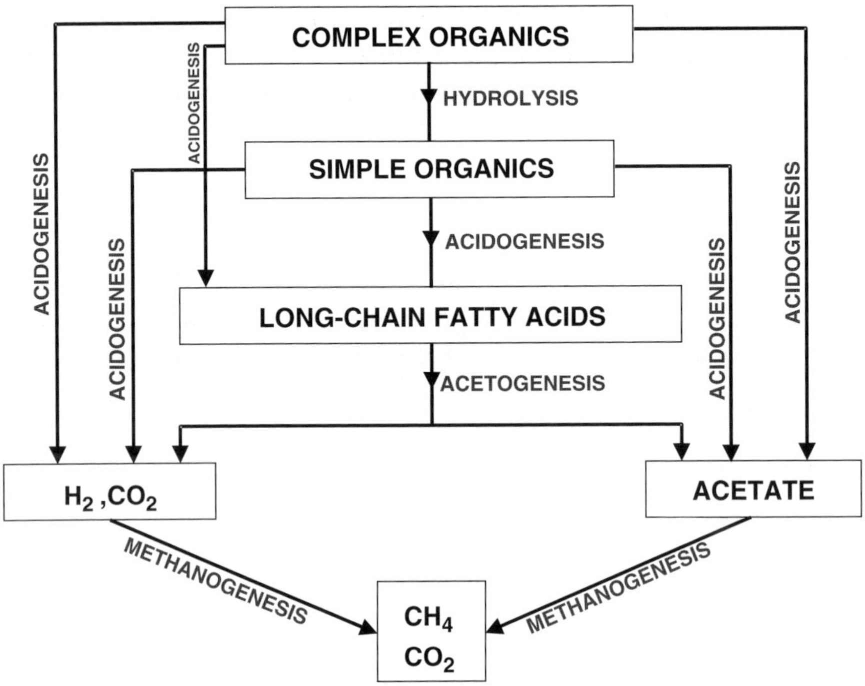

Fig. 1.1. Simplified pathways of methane fermentation of complex wastes.

steps involved due to the many pathways available for an anaerobic community. Figure 1.1 illustrates simplified pathways of methane fermentation of complex wastes by various routes. The microbial species including methanogens and acidogens form a syntrophic relationship in which each bacteria group constitutes a significant link in a complex chain of bioconversion. The syntrophic microcolony model suggests that the syntrophic relationship eventually lead to the formation of stable microcolonies or consortia, viz initial granules (Hirsh, 1984). Anaerobic granule indeed can be regarded as the congregation of cells to form fairly stable, contiguous, multicellular associations under physiological conditions in a defined biological system. The close packing of bacteria in granule architecture inherently facilitates the exchange of metabolites.

In UASB granules, different groups of bacteria carry out sequential metabolic processes, and interspecies syntrophic reactions are energetically beneficial. Because of the need for such close proximity, random cell-to-cell association in UASB granules would not enhance metabolic reactions. As pointed out by Fang (2000), "biogranules are developed through evolution instead of random aggregation of suspended microbes". In order to maintain high metabolic efficiency, the granule-associated cells would present in an organized structure, and signaling mechanisms in organizing the syntrophic species can be predicted (Shapiro, 1998). Therefore, it appears from the syntrophic microcolony model that the driving force for sludge granulation should be a result of the needs for bacterial survival or balance and for optimal combination of different biochemical functions of multiple species under the culture conditions.

Multilayer Model

Based on the microscopic observations, a multilayer model for anaerobic granulation was initially proposed by MacLeod et al. (1990); Guiot et al. (1992). According to this model, the microbiological composition of granules is different in each layer. The inner layer mainly consists of methanogens that may act as nucleation centers necessary for the initiation of granule development. H_2-producing and H_2-utilizing bacteria are dominant species in the middle layer, and a mixed species including rods, cocci, and filamentous bacteria takes predominant position in the outermost layer (Fig. 1.2). To convert a target organic to methane, the spatial

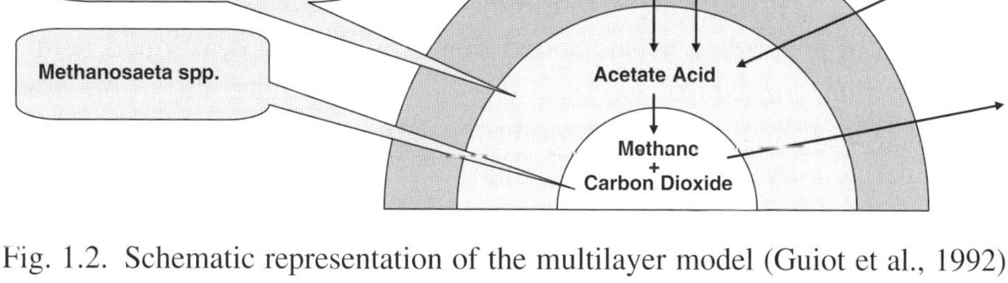

Fig. 1.2. Schematic representation of the multilayer model (Guiot et al., 1992).

organizations of methanogens and other species in UASB granules are essential.

The layered structure of UASB granules is supported by the works of Ahring et al. (1993); Lens et al. (1995) with immunological and histologic methods, with a dynamic model (Arcand et al., 1994), with microelectrodes (Santegoeds et al., 1999), and with fluorescence *in situ* hybridization using 16S rRNA-targeted oligonucleotides (Sekiguchi et al., 1998, 1999; Tagawa et al., 2000). A distinct layered structure was also found in the methanogenic–sulfidogenic aggregates, with sulfate-reducing bacteria in the outer 50–100 μm and methanogens in the inner layers (Santegoeds et al., 1999). Unlike the initial multilayer model proposed by MacLeod et al. (1990), recent research showed that UASB granules had large dark non-staining centers, in which neither archaeal nor bacterial signals could be found (Rocheleau et al., 1999). In fact, the non-staining center in the UASB granules might be formed as a result of the accumulation of metabolically inactive, decaying biomass, and inorganic materials (Sekiguchi et al., 1999).

Ecological Models

From microscopic examination and activity measurements, Dubourgier et al. (1987) suggested that granulation mechanism starts by the covering of filamentous Methanothrix by colonies of cocci or rods (acidogenic bacteria), forming microflocs of 10–50 μm. Subsequently, Methanothrix filaments, due to its filamentous morphology and surface properties, might establish bridges between several microflocs forming larger granules of size greater than 200 μm. Further development of acidogenic and syntrophic bacteria favors the granules growth. The authors support the idea that Methanothrix plays a vital role in enhancing granule strength by forming a network that stabilizes the overall structure. The role of extracellular polymers and cell walls are also emphasized.

Morgan et al. (1991a,b) suggested that granules are developed from a precursor that consists of a small aggregate of Methanothrix and other bacteria. Growth of the Methanothrix filaments form distinctive bundles separated by a surrounding matrix in which other methanogenic and non-methanogenic bacteria are embedded. As the bundles increase in size, the surrounding matrix is excluded leading to a region towards the center of the granule, which consists exclusively of compact filaments of Methanothrix and where discrete bundles are not distinguishable. Thus, the authors support previous suggestions on the importance of Methanothrix and bacterial polymers in the growth of the granules.

From the research developed in 1980s, de Zeeuw (1988) explains the formation of three types of granules developed in laboratory UASB reactor start-up experiments using digested sludge as inoculum and VFA as substrate. *Methanothrix* and *Methanosarcina* seem to be of predominant significance for granule formation. The characteristics of the formed granules were described as follows:

(A) Compact spherical granules mainly composed of rod-shaped bacteria resembling *Methanothrix soehngenii* in short chains or single cells (rod-granules).
(B) More or less spherical granules mainly consisting of loosely intertwined filamentous bacteria attached to an inert particle (filamentous granules). The prevailing bacteria resembled *Methanothrix soehngenii.*
(C) Compact spherical granules composed predominantly of *Methanosarcina*-type bacteria (Fig. 1.3).

The development of each type of granular sludge was explained on the basis of seed sludge selection and sludge bed erosion and expansion, and the consequent differences in selection pressure and mean sludge

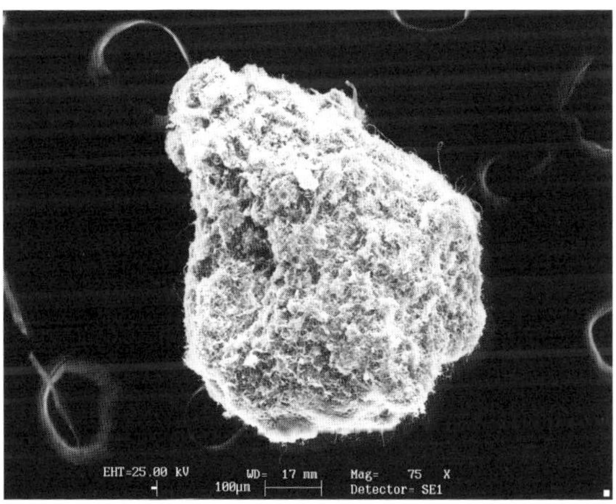

Fig. 1.3. Aggregate of *Methanosarcina* present at the bottom of a UASB reactor.

residence time. *Methanosarcina* granules develop due to the capacity of this genus to produce clumps of bacteria independently of the selection pressure. The clumps can reach macroscopic dimensions and show cavities, which can be inhabited by other species (Bochem et al., 1982). However, this kind of granules were just found in experiments where the concentration of acetate as a sole substrate was maintained above 1000 g COD/m^3, which means that *Methanosarcina* was able to outcompete *Methanothrix* (de Zeeuw, 1984).

At the low loading rates (low selection pressure) applied during the initial phase of a UASB reactor start-up, *Methanothrix* filaments will grow in and on small flocs present in the seed sludge leading to the formation of a "bulking" anaerobic sludge.

When a high selection pressure is applied, *Methanothrix*, that has a high affinity to attach to all kind of surfaces (van den Berg and Kennedy, 1981), attach onto carrier materials originating from the seed sludge or from the wastewater itself forming filamentous granules (type B).

More compact *Methanothrix* granules (rod granules, type A) are thought to be formed by the colonization of the central cavities of *Methanosarcina* clumps by *Methanothrix* bacteria, which have a higher acetate affinity, eventually leading to a loss of the outer layer of *Methanothrix*. Another explanation for these rod-type granules can be the filling of the filamentous granules with more bacteria, leading to a more compact Methanothrix granule.

The development of A or B type granules is related to the mean cell residence time maintained in the start-up process. When the mean cell residence time is too short, the opportunity to form compact granules consisting almost exclusively of biomass is slim. This means that large conglomerates of bacteria can only be formed through attachment to inert carriers, which must be heavy enough to be retained in the reactor (type B). Compact bacterial granules (type A) would only be formed if the mean cell residence time is sufficiently long.

Consideration on the Structural Models

Capetown model postulates that anaerobic granulation would not take place in UASB reactors treating acetate, propionate, or butyrate because of inadequate hydrogen partial pressure. However, there was experimental evidence that anaerobic granules could be formed in UASB systems fed with the substrates mentioned above (Ahring et al., 1993; van Lier et al., 1995). On the other hand, high hydrogen partial pressure is not desirable with respect to the granule-associated bacteria, because the partial pressure of hydrogen must be maintained at low level to ensure efficient fermentation of the volatile fatty acids. This may imply that the Capetown model is applicable only under some specific conditions. The importance of ECP in anaerobic granulation has been evidenced (Schmidt and Ahring, 1994, 1996). It seems that ECP may play an important role in building spatial structure and maintaining the stability of granules. However, the contribution ECP to the initiation of anaerobic granulation remains debatable. In addition, a high amount of ECP seems unnecessary for forming active granules. Instead, it has been found that too much ECP may deteriorate floc formation (Schmidt and Ahring, 1996).

In the syntrophic microcolony model, a close synergistic relationship among different microbial groups is essential for breaking down the complex organic wastes. Syntrophic microcolonies provide the kinetic and thermodynamic requirements for intermediate transference and therefore efficient substrate conversion (Schink and Thauer, 1988). It seems certain that the synergistic requirements provide a driving force for bacteria to form granules, in which different microbial species function in a synergistic way to increase the chance of survival.

Contrary to the multilayer model, anaerobic granules with non-layered structure were also reported (Grotenhuis et al., 1991; Fang et al., 1995; Wu et al., 2001). There was evidence that a layered and non-layered microstructure of the UASB granules may be developed with carbohydrates and substrates having a rate-limiting hydrolytic or fermentative step (e.g. proteins), respectively (Fang et al., 1995; Fang, 2000). This is probably due to different initial steps in the carbohydrate and protein degradation. The initial carbohydrate degradation to small molecules is faster than its subsequent degradation of the intermediates, whereas the initial step in the protein degradation is a rate-limiting step. Results from fluorescence *in situ* hybridization combined with confocal scanning laser microscopy showed that protein-fed granules possesses non-layered structure with a random distribution of *Methanosaeta concilii* (Rochclcau ct al., 1999). However, different types of granules may also form on the same substrate (Daffonchio et al., 1995; Schmidt and Ahring, 1996). Based on microscopic examination of the UASB granules, Fang (2000) proposed that the microbial distribution of the UASB granules strongly depends on the degradation thermodynamics and kinetics of individual substrates. Therefore, it appears that different dominating catabolic pathways may give rise to different structural granules.

So far, none of the structural models could explain a spontaneous and sudden washout of the established granular sludge bed as a result of a change in wastewater composition, which is commonly encountered in the operation of UASB systems. The point is, if a factor that is independent of the wastewater composition can initiate the formation of UASB granules, a change in the wastewater composition should not lead to the washout of the entire granular sludge bed. Thus, it is a reasonable speculation that there should be a substrate composition-associated factor that could also contribute to the formation of UASB granules. However, this proposition is yet to be included in the current structural models discussed previously.

Proton Translocation–Dehydration Theory

Theory Development

Several researchers observed the essential of proton translocation concept that (i) the hydrophobic interaction of a considerable extent was closely related to the initiation of bacterial adhesion; (ii) the proton conductance across a bacterial surface could induce surface dehydration; and (iii) the proton translocating activity could induce the protonation of bacterial cell surfaces. Based on these observations and a consideration of the proton translocating activity on bacterial membrane surfaces, a proton translocation–dehydration theory for molecular mechanism of sludge granulation was proposed and proved by experiments (Teo et al., 2000; Tay et al., 2000a). The theory suggests that the overall sludge granulation process in a typical anaerobic wastewater treatment system is initiated by the bacterial proton translocating activity at bacterial surfaces.

Dehydration of Bacterial Surfaces

During the start-up, the substrate is fed into an anaerobic reactor which has been inoculated with seed sludge. The fermentative bacteria secrete extracellular enzymes into the medium to catalyze the hydrolysis/acidification of the organic compounds. The compounds are degraded into volatile fatty acids coupling with the electron transport. Simultaneously, the proton pumps on the membranes of these bacteria are activated. The proton translocating activity can establish a proton gradient across the bacterial cell surface and subsequently cause surface protonation. The energized bacterial surfaces result in the breaking of hydrogen bonds between negatively charged groups and water molecules as well as partial neutralization of the negative charges on their surfaces. This in turn induces the dehydration of the bacterial surfaces.

Embryonic Granule Formation

The fermentation of complex organic compounds supplies the substrates to acetogens and methanogens and accelerates their growth and duplication. Similarly, coupling with the electron transport on their respiration chains, the acetogenic and methanogenic bacterial surfaces are dehydrated due to the presence of high-energy protons. By the action of external hydraulic forces, these relatively neutral and hydrophobic acidogens, acetogens, and methanogens may adhere to each other to form embryonic granules due to the weaker hydration repulsion. These initial aggregates are strengthened by further dehydration of the bacterial surfaces, which results from the effective metabolites transference. Only those embryonic granules that are able to obtain energy and nutrients from the environment are selected. Moreover, this new physiological environment begins to induce the excretion of extracellular polymers (ECPs) to the embryonic granule surfaces.

Granule Maturation

Within each embryonic granule, there is an on-going methanogenic series metabolism. Distribution of each group of bacteria in the granules depends on the orientation of intermediate metabolites transference, which is believed to be the most efficient way for anaerobes to transfer their intermediates. Formation of well-organized bacterial consortia as mature granules is thus possible. Embryonic granules may also adhere to and integrate other dispersed bacteria while the original bacterial colonies (or consortia) continue to grow and multiply. Granule maturation resists and blocks the unrestricted multiplication of bacterial cells because of space restriction for them to grow and to dispose off metabolites waste products. This space restriction and the continuous supply of substrates facilitates the production of ECP in large quantities.

Post-maturation

The bacterial proton translocating activity in mature granules keeps the bacterial surfaces at a relatively hydrophobic state. Maintenance of the structure of mature granules is governed by the mechanism of proton translocation–dehydration. On the other hand, an ECP outer layer causes the hydration of the granule surface, which protects the granule against attachment to gas bubbles and shear stress existing in the UASB reactor.

Consideration on the Proton Translocation–Dehydration Theory

The proton translocation-induced dehydration of bacterial surface is considered a key element of the proton translocation–dehydration theory. In accordance with the chemiosmotic mechanism on most of the aerobic bacteria, ATP is generated by oxidative phosphorylation, in which process electrons are transported through the electron transport system (ETS) from an electron donor (substrate) to a final electron acceptor (O_2). The molecules directly using the H^+ gradient built up by electron transport can be considered H^+-ATP as pumps. In anaerobic methanogens, ATP synthesis is linked with methanogenesis by electron transport, proton pumping, and a chemiosmotic mechanism (Prescott et al., 1999). Similar to aerobic respiration, anaerobic respiration is effective because it is more efficient than fermentation and allows ATP synthesis by electron transport and oxidative phosphorylation in the absence of oxygen. Thus, it appears that proton translocation-driven phosphorylation is a common mechanism for energy generation in both aerobic and anaerobic respirations. It should be pointed out that some bacteria, for example, *Streptococcus*, have no respiration chain and can produce ATP only via substrate-level phosphorylation. In this case, the proton gradients across those bacterial surfaces are often generated by proton extrusion catalyzed by membrane ATPases (H^+/ATPases) at the expense of ATP. It follows that the metabolic end-product efflux is an additional mechanism for proton extrusion from *Streptococci* and other bacterial cells that result in the generation of proton gradients. Protons are disposed off as acid to regulate their cytoplasmic pH conditions. This in turn can cause protonation and dehydration on the bacterial surfaces.

The fundamentals of energy metabolism show that proton translocation across cellular membrane exists in both aerobic and anaerobic respirations. It has been well established that anaerobic respiration is not as efficient as aerobic respiration in ATP synthesis, because the alternate electron acceptors, such as nitrate, sulfate, or carbon dioxide have less positive reduction potentials than oxygen (Prescott et al., 1999). This implies that less energy is available to generate ATP in anaerobic respiration. In other word, the proton translocation activity across cellular membrane in anaerobic respiration is much lower than that in aerobic process.

The proton translocation-induced dehydration theory suggests that microbial granulation could be observed in any aerobic or anaerobic system, and is independent of the types of substrate, bioreactors, and operation conditions. However, microbial granulation has never been reported in conventional activated sludge systems in the last 100 years of operation, and that anaerobic granules are formed mostly in UASB process. Feasibility and efficiency of other types of anaerobic bioreactors with development of anaerobic granules have not been sufficiently demonstrated yet. The proton translocation–dehydration theory provides useful information in understanding how anaerobic granules are developed in a molecular level. However, this theory does not account for those conditions-associated metabolic changes/requirements of microorganisms, which are considered significant contributors to the formation of UASB granules.

Cellular Automaton Model

Cellular automaton model has been used to describe the formation of microcolonies and biofilms (Ben-Jacob et al., 1991; Wimpenny and Colasanti, 1997; Kreft et al., 2001). The cellular automaton model is defined as spatially and temporally discrete system where the state of an automaton is determined by a set of rules that act locally but apply globally (Wimpenny and Colasanti, 1997). In the model, cellular automata form a class of systems composed of individual units (cells), each with a defined state, and each cell can change its state following the transition rules, which are influenced by its own state and those of other cells (Wimpenny and Colasanti, 1997). This model aims to reproduce a microbial structure under substrate-transfer-limited conditions. Substrate gradients created by local consumption of substrate allow the bacteria situated on "mounds" to have more substrate available than those situated in "valley" (Tolker-Nielsen and Molin, 2000). Thus, the structure of microcolony or biofilm is related to the availability of resource. Details of the automaton model have been described by Wimpenny and Colasanti (1997).

It had been reported that a simple and practical way towards rapid anaerobic granulation was to increase the organic loading rate based on an 80% reduction of biodegradable chemical oxygen demand with supplementary monitoring of effluent suspended solids washout (de Zeeuw, 1988; Fang and Chui, 1993; Tay and Yan, 1996). The findings are consistent with the prediction of the cellular automaton model which simulates a dynamic development of a microcolony or biofilm under varying environmental conditions. The model can in fact produce a large variety of distinct morphologies in response to changes in growth conditions (Ben-Jacob et al., 1991; Wimpenny and Colasanti, 1997). However, the cellular automaton model does not account for cell mobility towards resource and the role of cell-to-cell communication in the development of spatial organization of microcolony or biofilm, as pointed out by Tolker-Nielsen and Molin (2000).

Recently, based on the cellular automaton theory, a series of multidimensional biofilm models with heterogeneous biomass and substrate distribution in two or three dimensions have been developed (Hermanowicz, 1997; Noguera et al., 1999; Picioreanu et al., 1999, 2001; Kreft et al., 2001). In the multidimensional biofilm models, it is generally assumed that biofilm growth is due to the processes of diffusion, reaction, and growth including biomass growth, division, and spreading. Many studies suggested that the structure of granules is rather similar to the structure of biofilms (MacLeod et al., 1990; Schmidt and Ahring, 1996; Tolker-Nielsen and Molin, 2000), thus the multidimensional models used to explain the spatial organization of bacteria in biofilms could be applied to anaerobic granulation.

It should be pointed out that as models are getting more and more complex, model calibration becomes a challenging task. Without an adequate calibration, quantitative results generated from modeling may become meaningless. Therefore, future study needs to look into the applicability of the multidimensional biofilm models to accommodate anaerobic granulation process.

Cell-to-Cell Communication Model

Although mechanisms and models for anaerobic granulation are available abundantly in the literature, none of them could provide a complete description for anaerobic granulation process. Intercellular communication and multicellular coordination have been known as an effective way for bacteria to achieve an organized spatial structure. It has been shown that quorum sensing is a prominent example of social behavior in bacteria, as signal exchange among individual cells allows the entire population to choose an optimal way of interaction with the environment.

The cellular automaton model shows that biofilm structure is determined by localized substrate concentration (Wimpenny and Colasanti, 1997), however it has been found that a cell indeed can read its position in a concentration gradient of an extracellular signal factor, and to determine its developmental fate accordingly (Gurdon and Bourillot, 2001). Based on recent research findings on cell-to-cell communication (Davies et al., 1998; Pratt and Kolter, 1999; Ben-Jacob et al., 2000), it can be predicted that cell-to-cell signaling mechanisms are effective in developing anaerobic granules and organizing the spatial structure of granule-associated bacteria in response to environmental stresses. In fact, larger-scale organization had been observed in the distribution of distinct species and of distinct metabolic processes within the UASB granules (Shapiro, 1998).

A number of different groups of bacteria are involved in carrying out sequential metabolic processes in anaerobic granules. In order to efficiently utilize a target organics, the bacteria need to be spatially organized. As summarized by Shapiro (1998), the benefits of an organized microbial structure include more efficient proliferation; access to resource and niches that cannot be utilized by isolated cells; collective defense against antagonists that eliminate isolated cells, and optimization of population survival by differentiation into distinct cell types. These are strongly supported by experimental evidence that UASB granules are much more resistant than suspended sludge to toxicity of hydrogen sulfide, heavy metals, and aromatic pollutants in wastewater (Bae et al., 2000; Fang, 2000; Tay et al., 2000a,b).

It has been generally observed in UASB reactors that a change in wastewater composition could result in a washout of the granular sludge within a short period of time. This phenomenon can be reasonably explained by the cell-to-cell communication mechanism. As pointed out earlier, the bacteria in a UASB granule are not randomly distributed but rather organized to best meet the needs of each species for a defined organic substrate. In fact, spatial organization of UASB granules is developed to cope with the constraints imposed by the substrate and corresponding metabolic processes. When the composition of wastewater is changed, the granule-associated bacteria would respond by re-organizing microbial spatial distribution and structure, in order to adapt to new metabolic processes required for the oxidation of present organic substrate.

Structure changes induced by a substrate shift have been reported in biofilm culture processes (Wolfaardt et al., 1994; Tolker-Nielsen and Molin, 2000). The substrate change-induced structural re-organization would result in a partial or complete breakup of the granules developed from the previous substrate. The observed washout of sludge blanket from UASB reactors is thought to be resulted from the substrate change-caused granule breakup. It appears from the cell-to-cell communication model that organized bacterial community, such as biofilms or granules, is not simply a scaled-up version of individual bacteria. Further research is required to refine the cell-to-cell communication-based mechanism for anaerobic granulation.

A General Model for Anaerobic Granulation

For bacteria in an anaerobic culture to form granules, a number of conditions have to be fulfilled. The contributions of physical, chemical, and biological factors to granulation process could not be considered separately. So far, no model seems able to depict the entire anaerobic granulation process reasonably. Based on the existing mechanisms for formation of anaerobic granules, a general four-step model to better describe anaerobic granulation is proposed as follows.

Step 1: Physical movement to initiate bacterium-to-bacterium contact or bacterial attachment onto nuclei. The forces involved in this step include:

- Hydrodynamic force.
- Diffusion force.
- Gravity force.
- Thermodynamic forces, e.g. Brownian movement.
- Cell mobility. Cells can move by means of flagella, cilia, and pseudopods, while cell movement may also be directed by a signaling mechanism.

Step 2: Initial attractive physical, chemical, and biochemical forces to keep stable multicellular contacts. These attractive forces are:

Physical forces:

- van der Waals forces.
- Opposite charge attraction.
- Thermodynamic forces including free energy of surface; surface tension.
- Hydrophobicity.
- Filamentous bacteria that can serve as a bridge to link or grasp individual cells together.

It should be emphasized that the hydrophobicity of bacterial surface plays a crucial role in the initiation of biofilms and anaerobic granules (Mahoney et al., 1987; van Loosdrecht et al., 1987; Teo et al., 2000; Tay et al., 2000a). According to the thermodynamics theory, increasing the hydrophobicity of cellular surfaces would cause a corresponding decrease in the excess Gibbs energy of the surface, which in turn promotes cell-to-cell interaction and further serves as a driving force for bacteria to self-aggregate out of liquid phase (hydrophilic phase).

Chemical forces:

- Hydrogen liaison.
- Formation of ionic pairs.

- Formation of ionic triplet.
- Interparticulate bridge and so on.

Biochemical forces:

- Cellular surface dehydration.
- Cellular membrane fusion.
- Signaling and collective action in bacterial community.

As described by the proton translocation–dehydration theory (Teo et al., 2000; Tay et al., 2000a), cellular surface dehydration and membrane fusion could lead to initiation of anaerobic granulation, while cooperative self-organization of bacteria will assist to form an organized spatial structure (Shapiro, 1998; Ben-Jacob et al., 2000).

Step 3: Microbial forces to make cell aggregation mature:

- Production of extracellular polymer by bacteria, such as exopolysaccharides.
- Growth of cellular cluster.
- Metabolic change and genetic competence induced by environment, which facilitate the cell–cell interaction and result in a highly organized microbial structure.

Step 4: Steady-state three-dimensional structure of microbial aggregate shaped by hydrodynamic shear forces. The microbial aggregates are finally shaped by hydrodynamic shear force to form a certain structured community. The shape and size of microbial aggregates are predominantly determined by the interactive strength/pattern between aggregates and hydrodynamic shear force, microbial species, and substrate loading rate.

The present general four-step model for anaerobic granulation attempts to broadly cover the current understanding of the entire granulation process as much as possible. It should be realized that identification of gross engineering events in relation to anaerobic granulation is relatively easy. But to identify the events at molecular or genetic level, a more profound understanding of the mechanisms responsible for anaerobic granulation is required. As Tolker-Nielsen and Molin (2000) noted, "it probably does not make sense to make firm decisions about one or the other explanation as the rule for community development".

References

Ahring, B.K., Schmidt, J.E., Winther-Nielsen, M., Macario, A.J.L., & de Macario, E.C. (1993). Effect of medium composition and sludge removal on the production, composition and architecture of thermophilic (55°C) acetate-utilizing granules from an upflow anaerobic sludge blanket reactor. *Appl. Environ. Microbiol.*, *59*, 2538–2545.

Alphenaar, P.A., Visser, A., & Lettinga, G. (1993). The effect of liquid upflow velocity and hydraulic retention time on granulation in UASB reactors treating wastewater with a high-sulphate content. *Bioresour. Technol.*, *43*, 249–258.

Arcand, Y., Guitot, S.R., Desrochers, M., & Chavarie, C. (1994). Impact of the reactor hydrodynamics and organic loading on the size and activity of anaerobic granules. *Chem. Eng. J. Biochem. Eng. J.*, *56*, 23–35.

Bae, J.W., Rhee, S.K., Hyun, S.H., Kim, I.S., & Lee, S.T. (2000). Layered structure of granules in upflow anaerobic sludge blanket reactor gives microbial populations resistance to metal ions. *Biotechnol. Lett.*, *22*, 1935–1940.

Ben-Jacob, E., Cohen, I., & Levine, H. (2000). Cooperative self-organization of microorganisms. *Adv. Phys.*, *49*, 395–554.

Ben-Jacob, E., Schochet, O., Tenenbaum, A., Cohen, I., Czirok, A., & Tamas, V. (1991). Generic modeling of cooperative growth patterns in bacterial colonies. *Nature*, *368*, 46–49.

Bochem, H.P., Schoberth, S.M., Sprey, B., & Wengler, P. (1982). Thermophilic biomethanation of acetic acid: morphology and ultrastructure of a granular consortium. *Canad. J. Microbiol.*, *28*, 500–510.

Cammarota, M.C., & Sant'Anna Jr., G.L. (1998). Metabolic blocking of exopolysaccharides synthesis: effects on microbial adhesion and biofilm accumulation. *Biotechnol. Lett.*, *20*, 1–4.

Chen, J., & Lun S.Y. (1993). Study on mechanism of anaerobic sludge granulation in UASB reactors. *Water Sci. Technol.*, *28*, 171–178.

Daffonchio, D., Thavessri, J., & Verstraete, W. (1995). Contact angle measurement and cell hydrohpobicity of granular sludge from upflow anaerobic sludge bed reactors. *Appl. Environ. Microbiol.*, *61*, 3676–3680.

Davies, D.G., Parsek, M.R., Pearson, J.P., Iglewski, B.H., Costerton, J.W., & Greenberg, E.P. (1998). The involvement of cell-to-cell signals in the development of a bacterial biofilm. *Science*, *280*, 295–298.

de Zeeuw, W.J. (1984). Acclimatization of anaerobic sludge for UASB reactor start-up. *Ph.D. Thesis*. Agricultural University Wageningen, The Netherlands.

de Zeeuw, W.J. (1988). Granular sludge in UASB-reactors. *Granular Anaerobic Sludge: Microbiology and Technology* (eds. Lettinga, G., Zehnder, A.J.B., Grotenhuis, J.T.C., & Hulshoff Pol, L.W.), Wageningen, The Netherlands, 132–145.

Dubourgier, H.C., Prensier, G., & Albagnac, G. (1987). Structure and microbial activities of granular anaerobic sludge. *Granular Anaerobic Sludge: Microbiology and Technology* (eds. Lettinga, G., Zehnder, A.J.B., Grotenhuis, J.T.C., & Hulshoff Pol, L.W.), Pudoc Wageningen, The Netherlands, 18–33.

El-Mamouni, R., Leduc, R., & Guiot, S.R. (1998). Influence of synthetic and natural polymers on the anaerobic granulation process. *Water Sci. Technol.*, *38*, 341–347.

Fang, H.H.P. (2000). Microbial distribution in UASB granules and its resulting effects. *Water Sci. Technol.*, *42*, 201–208.

Fang, H.H.P., & Chui, H.K. (1993). Maximum COD loading capacity in UASB reactors at 37°C. *J. Environ. Eng.*, *119*, 103–119.

Fang, H.H.P., Chui, H.K., & Li, Y.Y. (1995). Effect of degradation kinetics on the microstructure of anaerobic biogranules. *Water Sci. Technol.*, *32*, 165–172.

Forster, C.F., & Lewin, D.C. (1972). Polymer interaction at activated sludge surfaces. *Effl. Water. Treat. J.*, *12*, 520–525.

Grootaerd, H., Liessens, B., & Verstraete, W. (1997). Effects of directly soluble and fibrous rapidly acidifying chemical oxygen demand and reactor liquid surface tension on granulation and sludge-bed stability in upflow anaerobic sludge blanket reactors. *Appl. Microbiol. Biotechnol.*, *48*, 304–310.

Grotenhuis, J.T.C., van Lier, J.B., Plugge, C.M., Stams, A.J.M., & Zehnder, A.J.B. (1991). Effect of ethylene glycol-bis(β-aminoethylether)-N, N-tetraacetic acid (EGTA) on stability and activity of methanogenic granular sludge. *Appl. Microbiol. Biotechnol.*, *36*, 109–114.

Guiot, S.R., Gorur, S.S., Bourque, D., & Samson, R. (1988). Metal effect on microbial aggregation during upflow anaerobic sludge bed-filter (UBF) reactor start-up. *Granular Anaerobic Sludge: Microbiology and Technology* (eds. Lettinga, G., Zehnder, A.J.B., Grotenhuis, J.T.C., & Hulshoff Pol, L.W.), Wageningen, The Netherlands, 187–194.

Guiot, S.R., Pauss, A., & Costerton, J.W. (1992). A structured model of the anaerobic granules consortium. *Water Sci. Technol.*, *25*, 1–10.

Gurdon, J.B., & Bourillot, P.Y. (2001). Morphogen gradient interpretation. *Nature*, *413*, 797–803.

Hermanowicz, S.W. (1997). A model of two-dimensional biofilm morphology. *Water Sci. Technol.*, *37*, 219–222.

Hirsh, R. (1984). Microcolony formation and consortia. *Microbial Adhesion and Aggregation* (ed. K.C. Marshall), Springer, Berlin, 373–393.

Hulshoff Pol, L.W. (1989). The phenomenon of granulation of anaerobic sludge. *Ph.D. Thesis*. Agricultural University of Wageningen, The Netherlands.

Hulshoff Pol, L.W., Heijnekamp, K., & Lettinga, G. (1988). The selection pressure as a driving force behind the granulation of anaerobic sludge. *Granular Anaerobic Sludge: Microbiology and Technology* (eds. Lettinga, G., Zehnder, A.J.B., Grotenhuis, J.T.C., & Hulshoff Pol, L.W.), Wageningen, The Netherlands, 153–161.

Imai, T. (1997). Advanced start up of UASB reactors by adding of water absorbing polymer. *Water Sci. Technol.*, *36*, 399–406.

Kalogo, Y., Seka, A.M., & Verstraete, W. (2001). Enhancing the start-up of a UASB reactor treating domestic wastewater by adding a water extract of Moringa oleifera seeds. *Appl. Microbiol. Biotechnol.*, *55*, 651–664.

Kreft, J.U., Picioreanu, C., Wimpenny, J.W.T., & van Loosdrecht, M.C.M. (2001). Individual-based modeling of biofilms. *Microbiol.*, *147*, 2897–2912.

Lens, P., de Beer, D., Cronenberg, C., Ottengraf, S., & Verstraete, W. (1995). The suse of microsensors to determine distributions in UASB aggregates. *Water Sci. Technol.*, *31*, 273–280.

Lettinga, G., van Velsen, A.F.M., Hobma, S.W., de Zeeuw, W., & Klapwijk A. (1980). Use of the upflow sludge blanket (USB) reactor concept for biological waste water treatment especially for anaerobic treatment. *Biotechnol. Bioeng.*, *22*, 699–734.

MacLeod, F.A., Guiot, S.R., & Costerton, J.W. (1990). Layered structure of bacterial aggregates produced in an upflow anaerobic sludge bed and filter reactor. *Appl. Environ. Microbiol.*, *56*, 1598–1607.

Mahoney, E.M., Varangu, L.K., Cairns, W.L., Kosaric, N., & Murray, R.G.E. (1987). The effect of calcium on microbial aggregation during UASB reactor start-up. *Water Sci. Technol.*, *19*, 249–260.

Morgan, J.W., Evison, L.M., & Forster, C.F. (1991a). Internal architecture of anaerobic sludge granules. *J. Chem. Technol. Biotechnol.*, *50*, 211–226.

Morgan, J.W., Evison, L.M., & Forster, C.F. (1991b). Upflow sludge blanket reactors: the effect of bio-supplements on performance and granulation. *J. Chem. Technol. Biotechnol.*, *52*, 243–255.

Mozes, N., & Rouxhet, P.G. (1987). Methods for measuring hydrophobicity of microorganisms. *J. Microbiol. Methods*, *6*, 99–112.

Noguera, D.R., Pizarro, G., Stahl, D.A., & Rittmann, B.E. (1999). Simulations of multispecies biofilm development in three dimensions. *Water Sci. Technol.*, *39*, 123–130.

Noyola, A., & Mereno, G. (1994). Granulation production from raw waste activated sludge. *Water Sci. Technol.*, *30*, 339–346.

O'Flaherty, V., Lens, P.N., de Beer, D., & Colleran, E. (1997). Effect of feed composition and upflow velocity on aggregate characteristics in anaerobic upflow reactors. *Appl. Microbiol. Biotechnol.*, *47*, 102–107.

Palns, S.S., Loewenthal, R.E., Dold, P.L., & Marais, G.R. (1987). Hypothesis for pelletisation in upflow anaerobic sludge blanket reactor. *Water SA*, *13*, 69–80.

Papahadjopoulos, D., Nir, S., & Duzgunes, N. (1990). Molecular mechanisms of calcium-induced membrane fusion. *J. Bioenerg. Biomemb.*, *22*, 157–175.

Pereboom, J.H.F. (1994). Size distribution model for methanogenic granules from full scale UASB & IC reactors. *Water Sci. Technol.*, *30* (12), 211–221.

Picioreanu, C., van Loodrecht, M.C.M., & Heijnen, J.J. (1999). Discrete-differential modelling of biofilm structure. *Water Sci. Technol.*, *39*, 15–122.

Picioreanu, C., van Loodrecht, M.C.M., & Heijnen, J.J. (2001). Two-dimensional model of biofilm detachment caused by internal stress from liquid flow. *Biotechnol. Bioeng.*, *72*, 205–218.

Pratt, L.A., & Kolter, R. (1999). Genetic analysis of bacterial biofilm formation. *Curr. Opin. Microbiol.*, *2*, 598–603.

Prescott, L.M., Harley, J.P., & Klein, D.A. (1999). *Microbiology*. McGraw-Hill, New York.

Rocheleau, S., Greer, C.W., Lawrence, J.R., Cantin, C., Laramee, L., & Guiot, S.R. (1999). Differentiation of *Methanosaeta concilii* and *Methanosarcina barkeri* in anaerobic mesophilic granular sludge by in situ hybridization and confocal scanning laser microscopy. *Appl. Environ. Microbiol.*, *65*, 2222–2229.

Rouxhet, P.G., & Mozes, N. (1990). Physical chemistry of the interaction between attached microorganisms and their support. *Water Sci. Technol.*, *22*, 1–16.

Rudd, T., Sterritt, R.M., & Lester, J.N. (1984). Complexation of heavy metals by extracellular polymers in the activated sludge process. *J. Water Pollut. Control. Fed.*, *56*, 1260–1268.

Sam-Soon, P.A., Looewenthal, R.E., Dold, P.L., & Marais, D.V.R. (1988). Pelletization in upflow anaerobic sludge bed reactors. *Anaerobic Digestion* (eds. Hall, E.R., & Hobson, P.N.), Pergamon Press, Oxford, UK, 55–60.

Sanchez, J.M., Arijo, S., Munoz, M.A., Morinigo, M.A., & Borrego, J.J. (1994). Microbial colonization of different support materials used to enhance the methanogenic process. *Appl. Microbiol. Biotechnol.*, *41*, 480–486.

Santegoeds, C.M., Damagaad, L.R., Hesselink, C., Zopfi, J., Lens, P., Muyzer, G., & de Beer, D. (1999). Distribution of sulfate-reducing and methanogenic bacteria in anaerobic aggregates determined by microsensor and molecular analysis. *Appl. Environ. Microbiol., 65*, 4618–4629.

Schink, B. & Thauer, R. (1988). Energetics of syntrophic methane formation and the influence of aggregation. *Proceedings of the Granular Anaerobic Sludge*, Pudoc, Wageningen, The Netherlands, 5–17.

Schmidt, J.E., & Ahring, B.K. (1993). Effects of magnesium on thermophilic acetate-degrading granules in upflow anaerobic sludge blanket (UASB) reactor. *Enzyme Microb. Technol., 15*, 304–310.

Schmidt, J.E., & Ahring, B.K. (1994). Extracellular polymers in granular sludge from different upflow anaerobic sludge blanket (UASB) reactors. *Appl. Microbiol. Biotechnol., 42*, 457–462.

Schmidt, J.E., & Ahring, B.K. (1996). Granular sludge formation in upflow anaerobic sludge blanket (UASB) reactors. *Biotechnol. Bioeng., 49*, 229–246.

Sekiguchi, Y., Kamagata, Y., Nakamura, K., Ohashi, A., & Harada, H. (1999). Fluorescence in situ hybridization using 16S rRNA-targeted oligonucleotides reveals localization of methanogenes and selected uncultured bacteria in mesophilic and thermophilic sludge granules. *Appl. Environ. Microbiol., 65*, 1280–1288.

Sekiguchi, Y., Kamagata, Y., Syutsubo, K., Ohashi, A., Harada, H., & Nakamura, K. (1998). Diversity of mesophilic and thermophilic granular sludge determined by 16S rRNA gene analysis. *Microbiol., 22*, 2655–2665.

Shapiro, J.A. (1998). Thing about bacterial populations as multicellular organisms. *Annu. Rev. Microbiol., 52*, 81–104.

Shen, C.F., Kosaric, N., & Blaszczyk, R. (1993). The effect of selected heavy metals (Ni, Co and Fe) on anerobic granules and their extracellular polymeric substance (EPS). *J. Water Res., 27*, 25–33.

Show, K.Y., Wang, Y. Foong, S.F., & Tay, J.H. (2004). Accelerated start-up and enhanced granulation in UASB reactors. *J. Water Res., 38* (9), 2293–2304.

Tagawa, T., Syutsubo, K., Sekiguchil, Y., Ohashi, A., & Harada, H. (2000). Quantification of methanogen cell density in anaerobic granular sludge consortia by fluorescence in-situ hybridization. *Water Sci. Technol., 42*, 77–82.

Tay, J.H., & Yan, Y.G. (1996). Influence of substrate concentration on microbial selection and granulation during start-up of upflow anaerobic sludge blanket reactors. *Water Environ. Res., 68*, 1140–1150.

Tay, J.H., Xu, H.L., & Teo, K.C. (2000a). Molecular mechanism of granulation. I:H^+ translocation-dehydration theory. *J. Environ. Eng., 126*, 403–410.

Tay, J.H., He, Y.X., & Yan, Y.G. (2000b). Anaerobic biogranulation using phenol as the sole carbon source. *Water Environ. Res., 72*, 189–194.

Teo, K.C., Xu, H.L., & Tay, J.H. (2000). Molecular mechanism of granulation—II: proton translocating activity. *J. Environ. Eng.*, *126*, 411–418.

Thaveesri, J., Daffonchio, D., Lessens, B., Vandermeren, P., & Verstraete, W. (1995). Granulation and sludge bed stability in upflow anaerobic sludge bed reactors in relation to surface thermodynamics. *Appl. Environ. Microbiol.*, *61*, 3681–3686.

Thiele, J.H., Wu, W.M., Jain, M.K., & Zeikus, J.G. (1990). Ecoengineering high rate biomathanation system: design of improved syntrophic biomathanation catalysis. *Biotechnol. Bioeng.*, *35*, 990–999.

Tolker-Nielsen, T., & Molin, S. (2000). Spatial organization of microbial biofilm communities. *Microb. Ecol.*, *40*, 75–84.

van den Berg, L., & Kennedy, K.J. (1981). Support materials for stationary fixed film reactors for high-rate methanogenic fermentations. *Biotechnol. Lett.*, *3*, 165–170.

van Lier, J.B., Sanx Martin, J.L., & Lettinga, G. (1995). Effect of temperature on the anaerobic thermophilic conversion of volatile fatty acids by dispersed and granular sludge. *J. Water Res.*, *30*, 199–207.

van Loosdrecht, M.C.M., Lyklema, J., Norde, W., Schraa, G., & Zehnder, A.J.B. (1987). Electrophoretic mobility and hydrophobicity as a measure to predict the initial steps of bacterial adhesion. *Appl. Environ. Microbiol.*, *53*, 1898–1901.

Wang, Y., Show, K.Y., Tay, J.H., & Sim, K.H. (2004). Effects of cationic polymer on start-up and granulation in UASB reactors. *J. Chem. Technol. Biotechnol.*, *79* (3), 219–228.

Wiegant, W.M. (1998). The Spaghetti theory on anaerobic granular sludge fermentation, or the inevitability of granulation. *Proceeding of the Granular Anaerobic Sludge*, Pudoc, Wageningen, The Netherlands, 146–152.

Wilschut, J., & Hoekstra, D. (1984). Membrane fusion: from liposome to biological membrane. *Trend Biochem. Sci.*, *9*, 479–483.

Wimpenny, J.W.T., & Colasanti, R. (1997). A unifying hypothesis for the structure of microbial biofilms based on cellular automaton models. *FEMS Microbiol. Ecol.*, *22*, 1–16.

Wolfaardt, G.M., Lawrence, J.R., Robarts, R.D., Caldwell, S.J., & Caldwell, D.E. (1994). Multicellular organization in degradative biofilm community. *Appl. Environ. Microbiol.*, *60*, 434–446.

Wu, W.M., Kickey, R.F., & Zeikus, J.G. (1991). Characterization of metabolic performance of methanogenic granules treating brewery wastewater: role of sulfate-reducing bacteria. *Appl. Environ. Microbiol.*, *57*, 3438–3449.

Wu, W.M., Jain, M.K., & Zeikus, J.G. (1996). Formation of fatty acid-degrading anaerobic granules by defined species. *Appl. Environ. Microbiol.*, *62*, 2037–2044.

Wu, J.H., Liu, W.T., Tseng, I.C., & Cheng, S.S. (2001). Characterization of microbial consortia in a terephthalate-degrading anaerobic granular sludge system. *Microbiol.*, *147*, 373–382.

Yu, H.Q., Tay, J.H., & Fang, H.H.P. (2001a). The role of calcium in sludge granulation during UASB reactor start-up. *J. Water Res.*, *35*, 1052–1060.

Yu, H.Q., Fang, H.H.P., & Tay, J.H. (2001b). Enhanced sludge granulation in upflow anaerobic sludge blanket (UASB) reactors by aluminum chloride. *Chemosphere*, *44*, 31–36.

Zita, A., & Hermansson, M. (1994). Effects of ionic strength on bacterial adhesion and stability of flocs in a wastewater activated sludge system. *Appl. Environ. Microbiol.*, *60*, 3041–3048.

Chapter 2

Factors Affecting Anaerobic Granulation

Kuan-Yeow Show

Introduction

A major problem associated with the upflow anaerobic sludge blanket (UASB) reactors is the long start-up period required for the development of anaerobic granules. In cases where the inoculation is done with municipal digester flocculant sludge, it usually takes 3 to 4 months or even a much longer period before the process can be put in operation. In view of the longer start-up period, enhanced granules formation is highly desirable in order to reduce space–time requirements of various bioreactors leading to cheaper treatment of high-strength wastes. The improvements can also lead to better treatment efficiency with greater capacity to handle large volumes of wastewater with more compact reactor design. It is therefore possible to economize on the capital investment and subsequent cost of daily operation.

Use of granular sludge from in-operating UASB reactors as the seed material has the advantage of being able to achieve high organics removal within a short start-up period. However, the availability of granular seed sludge is limited, and the costs for purchase and transportation of the inoculum are extremely high (approximately US$ 500–1000 per ton wet weight) (Liu et al., 2002). Consequently, technologies for enhanced and rapid production of anaerobic granules are sought after. While the approach for rapid production of anaerobic granules is being improved, a review of information on the selection pressure influencing anaerobic granulation can serve as a useful reference and guide. The selection pressure may result from environmental conditions (e.g. temperature, pH, and feed), process operating conditions (e.g. hydraulic retention time, upflow liquid velocity, characteristics of seed and substrate, and organic loading rate), and chemical conditions (e.g. effects of cations and polymers).

Environmental Conditions

Temperature

Anaerobic decomposition of organics is accomplished through a series of biochemical reactions which is very dependent on temperature. Most take place at mesophilic condition. As a core microbial component of anaerobic granules, methanogenic bacteria grow slowly in wastewater and their generation times range from 3 days at 35°C to as high as 50 days at 10°C (Bitton, 1999). When the reactor temperature is below 30°C, the activity of methanogens is seriously reduced. Although high temperature seems to increase the pace of granulation, most bacteria will lose their activity if the temperature is too high. Experiments showed that if temperature is increased suddenly from 35° to 55°C, sludge washout and lower COD removal efficiency was observed (Fang and Lau, 1996); Lepisto and Rintala (1999) further reported that effluent quality from a UASB reactor operated at 70°C was lower than that from reactors operated at 35 and 55°C.

There is an optimum range of temperature for successful functioning of anaerobic system. Most UASB reactors are operated at mesophilic range though some can be operated at a temperature as high as 70°C. However, there is seemingly no advantage to operate a UASB reactor at such a high temperature when the reactor can operate well at 35°C. High temperatures are known to encourage the growth of suspended biosolids; however, extremely high temperatures inhibit bacterial growth. Extreme thermophilic UASB reactors (i.e. temperature above 55°C) are impracticable because of the additional energy required to maintain the high temperature and the relatively poor effluent quality. This is indeed the main reason why mesophilic UASB reactors are more attractive as compared to their thermophilic counterparts. Moreover, a high-temperature operation is also difficult to control.

Recently, attention has been given to the impact of low temperature on the performance of anaerobic granular sludge reactors (Angenent et al., 2001; Lettinga et al., 2001; Lew et al., 2003; Singh and Viraraghavan, 2003). Singh and Viraraghavan (2003) showed that COD removal efficiency can be as high as 70 to 90% in a UASB reactor operated at 11°C with a hydraulic retention time of 6 h. Similarly, the expanded granular sludge bed (EGSB) reactors have been shown to be practicable systems for anaerobic treatment of mainly soluble and pre-acidified wastewaters at temperatures of 5 to 10°C (Lettinga et al., 2001). In addition, anaerobic migrating blanket reactors (AMBRs) have also been successfully applied to treat low-strength wastewaters at low temperatures (Angenent et al., 2001). Therefore, it is clear that anaerobic granular sludge systems are most suitably operated for the treatment of municipal wastewater at low and moderate temperatures.

System pH

Based on the sequence of anaerobic reaction, microbial species involved can be roughly divided into the following three categories: (a) bacteria responsible for hydrolysis; (b) acid-producing bacteria; and (c) methane-producing bacteria. In general, the acid-producing bacteria tolerate a low pH with an optimal pH of 5.0 to 6.0; however, most methane-producing bacteria can only function optimally in a very narrow pH range of 6.7–7.4 (Bitton, 1999). This explains why pH is more inhibitory to methane-producing bacteria than to acidogenic bacteria in UASB reactors. Once the reactor pH falls outside the range of 6.0–8.0, the activity of methane-producing bacteria is adversely affected which poses serious operational problem leading to reactor failure. Under normal operating conditions, the pH reduction caused by acid-producing bacteria can be buffered by bicarbonate produced by the methane-producing bacteria.

Teo et al. (2000) studied the effects of the environmental pH on anaerobic granulation process. They found that from pH 8.5 to 11.0, the strength of anaerobic granules in term of turbidity change decreased with the pH increase, indicating that high pH conditions weakened the granular structure; from pH 5.5 to 8.0, the strength of granules was unchanged, showing that the granular structure was relatively stable at this pH range; from pH 3.0 to 5.0, the increase in the strength of granule was very sharp. These results showed that the relatively low pH conditions would facilitate the maintenance of anaerobic granular structure, and can be satisfactorily explained by the proton translocation–dehydration theory. Consequently, *in situ* operation engineers need to regularly monitor the reactor pH and its changes.

Characteristics of the Feed

Characteristics of the feed are considered a key factor influencing the formation, composition, and structure of anaerobic granules. The complexity of substrate may exert a selection pressure on microbial diversity in anaerobic granules which influences the formation and microstructure of granules. Based on their free energy of oxidation, organic substrates can be roughly classified into high-energy and low-energy feeds. During the UASB start-up period, high-energy carbohydrate feeding can sustain the acidogens and facilitate the formation of extracellular polymers. The more readily the acidogens take up and metabolize the substrate, the more rapidly the proton pumps will be activated, and sooner the methanogens will obtain the substrate (Tay et al., 2000). Thus, the rapid growth of acidogens due to the presence of high-energy substrate in the influent would facilitate the overall process of sludge granulation in the UASB reactors.

The granules grown on volatile fatty acid mixture (acetate, propionate, and butyrate) under mesophilic conditions can be classified into three distinct types according to the predominant acetate utilizing methanogens present: (1) rod-type granules, which are mainly composed of rod-shaped bacteria in fragments of about four to five cells resembling *Methanothrix*; (2) filament-type granules, which consist predominantly of long multicellular rod-shaped bacteria; and (3) sarcina-type granules, which develop when a high concentration of acetic acid is maintained in the reactor (Hulshoff Pol et al., 1983; de Zeeuw, 1984).

A trend has been observed towards increasing diversity of methanogenic subpopulations with an increasing complexity of the waste composition. At least four distinct microcolonies have been observed in granules treating brewery wastewater (Wu, 1991). One of these microcolonies was composed of *Methanothrix*-like rods only, while the other microcolonies consisted of hydrogen–carbon dioxide utilizing *Methanobacterium*-like rods juxtapositioned with three different rod-shaped syntrophs (Hickey, 1991).

Full-scale UASB experience confirms that anaerobic sludge granulation occurs in many different types of wastewaters. Because of the extremely low growth rate of anaerobic bacteria, the energy content of the substrate are important for anaerobic granulation; however, the complexity of substrate also exerts a selection pressure on the microbial diversity in anaerobic granules. This selection pressure may in turn influence the formation and microstructure of granules through its effect on the food chain and community signaling communications.

Process Conditions During Start-up and Operation

Upflow Velocity and Hydraulic Retention Time

In a UASB reactor, upflow velocity and hydraulic retention time (HRT) is inter-related and serves as a selection pressure on microbial ecology. It has been observed that anaerobic granulation can proceed well at relatively high liquid upflow velocity, but does not occur under conditions of low hydrodynamic shear (Alphenaar et al., 1993; Arcand et al., 1994; O'Flaherty et al., 1997; Alves et al., 2000). According to Alphenaar et al. (1993), granulation in UASB reactors is favored by a combination of high upflow liquid velocity and short hydraulic retention time. Usually, the effects of upflow liquid velocity on anaerobic granulation are explained by the selection pressure theory (Hulshoff Pol et al., 1988). A long HRT accompanied with a low upflow liquid velocity may allow dispersed bacterial growth and be less favorable for microbe granulation. In contrast, a short HRT in association with a high upflow liquid velocity can lead to washout of flocculant biological solids and thus promotes sludge granulation.

Research attempts have been given to develop strategy for speed-up of granulation process by controlling hydrodynamic conditions in a UASB reactor. Noyola and Mereno (1994) conducted a series of experiments to investigate the effect of liquid upflow velocities for a rapid formation of granules through a purely physical aggregation due to the hydraulic stress applied to the anaerobic flocculant sludge with increasing upflow velocities. Experiments showed that flocculant anaerobic sludge could be converted to a relatively active anaerobic granular sludge by enhancing agglomeration with only hydraulic stress in a very short time less than 8 h, and the settleability of those anaerobic granules in terms of sludge volume index (SVI) and sludge settling velocity were significantly improved as the liquid upflow velocity increased. The increased settleability of granules in turn reduced washout of sludge from 46 to 2%.

Similarly, Arcand et al. (1994) also reported that the liquid upflow velocity had a significant positive effect on mean granule size, but its effect on the specific washout rate of the smaller particles was little. It is most likely that relatively high upflow velocity combined with a short HRT seem to be in favor of fast formation and production of anaerobic granular sludge. However, for a successful start-up and stable operation of UASB reactors, the reactor HRT would not be below a critical value, namely the minimum HRT.

Organic Loading Rate (OLR)

The OLR is related to the amount of "food" available for bacteria growth. In a microbiological sense, the OLR describes the degree of starvation of the microorganisms in a biological system. At a low OLR, microorganisms are subject to nutrient starvation, while a high OLR sustains fast microbial growth (Bitton, 1999). Research efforts have been dedicated to the role of organic loading rate (OLR), which is one of the most important operating parameters in anaerobic granulation process.

Evidence shows that anaerobic granulation can be accomplished by gradually increasing the OLR during the start-up (Hulshoff Pol, 1989; Kosaric et al., 1990; Campos and Anderson, 1992; Tay and Yan, 1996). It is critical to select a reasonably high OLR during start-up, to ensure rapid granulation and a stable treatment process. A simple and practical strategy for rapid start-up of anaerobic granular sludge reactors is to increase the OLR to attain only 80% reduction of biodegradable chemical oxygen demand (COD) with supplementary monitoring of effluent for washout of suspended solids (de Zeeuw, 1988; Fang and Chui, 1993).

An unconventional approach to accelerate start-up and granulation processes in UASB reactors has been developed by stressing the organic loading rate (OLR) without having to reach steady-state conditions (Show et al., 2004). The results indicate that the start-up of reactors could be significantly accelerated under stressed loading conditions. Start-up times of the moderately and severely stressed reactors for operating at OLRs of 1 to 16 g COD/l.d ranged from 10 to 80 days and 13 to 90 days, respectively. Comparing with 17 to 120 days needed in the control reactor to reach the same OLRs, the start-up times were shortened by 25 to 41%. The extent of acceleration depends on the level at which the reactor are stressed. Applying stress and the extent of stress level in starting up the reactors did not reduce the reactor loading capacity, as all the reactors reached a similar maximum OLR of 16 g/l.d at the end of operation.

Development of granulation could be accelerated with the unconventional approach of stressed loadings as demonstrated by the results. Under stressed loading conditions, the sludge particles began to form granules earlier in both the stressed reactors after 24 and 30 days of start-up operation. Comparing with the control reactor without applying stress, the times taken to form granule were reduced by 45 and 32% in the severely and moderately stressed units, respectively. The granule formation occurred earlier in the severely stressed reactor than the moderately stressed unit.

While the results obtained had established significant acceleration in start-up and granulation processes, the characteristics of granules developed were greatly influenced by the level of stress exerted. Characterization of bioparticles revealed that the granules developed in the moderately stressed reactor exhibited superior characteristics in terms of settleability, strength, microbial activity and morphology, and granular sludge growth, as compared with both the control reactor operated without stress and the unit which was over-stressed.

Tay and Yan (1996) further proposed that the start-up operation of UASB reactors could be guided by a dimensionless parameter, namely microbial load index (MLI). The MLI is defined by the ratio of OLR applied to specific methanogenic activity (SMA) in terms of gram methane-COD produced by gram VSS per day. An MLI value of around 0.8 was proved appropriate for rapid UASB start-up and microbial granulation. It should be pointed out that the MLI indeed is proportionally related to OLR, i.e. the MLI represents the magnitude of OLR. Large *Methanothrix*-like species (thrix granules) were cultivated with 1000 to 5000 mg COD/l influent, and small *Methanosarcina*-like species (sarcina granules) were cultivated with 10,000 mg COD/l influent. The thrix granules with median diameters of 2.5 to 3.4 mm exhibited better settleability, higher substrate affinity, and slightly higher bioactivity than the 0.54 mm sarcina granules (Tay and Yan, 1996).

The OLR-associated negative effects have been observed in UASB operation practice. High OLR results in a reduced mechanical strength of granules, i.e. the granules would easily lose their structural integrity, and disintegration would occur (Quarmby and Forster, 1995). Increased biogas production accompanied with high OLR would eventually lead to disintegrated granular sludge being washed out from the reactor. When the best-known Monod model is applied to the UASB system, an increased OLR will raise proportionally the biomass growth rate (Morvai et al., 1992). High growth rate of microorganisms would reduce the strength of three-dimensional structure of microbial community. Such a phenomenon has been observed in biofilm reactors (Liu and Tay, 2001). On the other hand, biogas production is also proportional to the magnitude of the applied OLR. If the applied OLR is too high in the period of start-up of UASB reactor, increased biogas production rate would cause serious hydrodynamic turbulence and further leads to the washout of seed sludge from the reactor, which sometimes is a main reason of unsuccessful start-up of UASB reactor. Table 2.1 shows some typical OLR values commonly used during the start-up of anaerobic granulation process, which provides some useful information on the OLR applied for UASB start-up.

Table 2.1. Some OLR values used for rapid start-up of UASB reactors (updated from Morvai et al., 1992)

Substrate	OLR at start-up (kg COD/kg VSS per d)	Time required for granulation (days)	References
Propionate	0.9	56–100	Hulshoff Pol et al. (1983)
Acetate	0.3	Not observed	Hulshoff Pol et al. (1983)
Brewery wastewater	0.28–0.63	41–40	Wu et al. (1985)
Sucrose wastewater	0.07–0.4	130–160	Wu et al. (1985)
Molasses wastewater	0.5–0.6	33–45	Wu et al. (1987)
Sucrose wastewater	0.1–0.38	36–70	Sierra-Alvarez et al. (1988)
Carbohydrate	0.4–1.2	28–45	Morvai et al. (1992)
Molasses wastewater	0.4–1.2	23–37	Morvai et al. (1992)
Synthetic wastewater	0.12	42–83	Campos and Anderson (1992)
Sucrose	0.2	21	Ghangrekar et al. (1996)
Sucrose	0.6	45	Ghangrekar et al. (1996)

Characteristics of Seed Sludge

Theoretically any medium containing the proper bacterial flora can be used as seed sludge for granule cultivation. Common seed materials include manure, fresh water sediments, septic tank sludge, digested sewage sludge, and surplus sludge from anaerobic treatment plants. Apart from its availability and its cost, the quality of a particular seed material can be judged in terms of ash content, the specific methanogenic activity, and the settleability.

Aerobic activated sludge from a sewage treatment plant and primary sludge from an aerobic plant treating textile dyeing wastewater had been used (Wu et al., 1987). It was found that there were sufficient anaerobic nuclei present in the aerobic flocs. All important methanogens seem to be present in aerobic activated sludge. Existing granules can also become seeding alternatives. Quality of seed sludge with respect to specific activity, settleability, and nature of inert fraction are important for anaerobic granulation process.

Two different types of sludges may develop on the same medium depending on the source of the inoculum. Xu and Tay (2002) used methanol-precultured anaerobic sludge to inoculate a UASB reactor. This approach accelerated the formation of embryonic granules in a laboratory-scale UASB reactor. The granulation process reached its postmaturation stage about 15 to 20 days ahead of the control reactor.

In engineering sense, heavy and relatively inactive sludge was preferred over lighter, more active sludge because of expected differences in washout. de Zeeuw (1984) observed two types of sludge washout, i.e. erosion washout and sludge bed expansion washout. Sludge bed erosion washout represents the selective washout on the basis of differences in settleability. Sludge bed expansion washout predominately occurs when using a diluted digested sewage sludge in the treatment of a medium strength wastewater. It is caused by the expansion of the sludge bed as a result of the increased hydraulic and gas loading rates and involves little selection between sludge particles with a difference in settleability. By choosing a concentrated digested sewage sludge as seed the latter type of sludge washout can be avoided.

Although digested sewage sludge is usually used for the start-up of a UASB reactor, various other types of seed sludge can be successfully utilized when granular sludge for seeding is unavailable. Wu et al. (1987) utilized aerobic activated sludge from a sewage treatment plant and primary sludge from an aerobic plant treating textile dyeing wastewater. Apparently, sufficient anaerobic nuclei were present in the aerobic flocs.

Using a MPN technique for counting the methanogens, it was found that aerobic activated sludge contains 10^8 methanogens/g suspended solids (SS), while in digested sludge, Zeikus (1979) found a number of 10^8/ml, giving for a 4% (w/v) sludge a figure of 2.5×10^{10}/g SS. All the important methanogens seem to be present in aerobic activated sludge. Other seed sludges that have been applied are lotus pond mud (Qi et al., 1985), cow manure (Wiegant, 1986), and primary sewage sludge (Ross, 1984).

The UASB system can also be started-up using existing granules whenever possible. This lends, in general, a decided advantage to the UASB process for start-up, although a successful start-up is not assured simply because granules are available. The inoculation with a large seed amount of granular sludge from a healthy UASB reactor is desirable. However, the availability of granular seed sludge is limited and the expenses for purchase and transportation of the inoculum are expensive. Addition of a small amount of granules to non-granular inoculum was still needed to stimulate the granulation process (Hulshoff Pol et al., 1983). This is probably a consequence of supplying an inoculum of microorganisms, which is responsible for granulation. On the other hand, Hulshoff Pol et al. (1983) reported that the addition of crushed granular methanogenic sludge to digested sewage in a UASB reactor fed with acetate plus propionate may give rise to the development of methanogenic sludge granules with a diameter of 1–2 mm. The observation that two different types of sludge developed on the same medium depending on the source of the inoculum, made in parallel experiments indicates that the formation of well settling conglomerates (i.e. granulation and pelletization) initially is a purely biological phenomenon.

The structures of anaerobic granules are closely related to the diversity of microorganisms. El-Mamouni et al. (1997) investigated the influence of four different granulation precursors, syntroph-enriched methanogenic consortia, *Methanothrix*-enriched, *Methanosarcina*-enriched nuclei, and acidogenic flocs on the development of anaerobic granules. It was found that granulation proceeded rapidly with syntroph-enriched methanogenic consortia, *Methanothrix*-enriched and *Methanosarcina*-enriched nuclei; however, granulation was significantly retarded when acidogenic flocs were used as precursors. The increase rate of granule size was 31 μm/day for syntroph-seeded granules, 21 μm/day for *Methanothrix*-seeded granules, 18 μm/day for *Methanosarcina*-seeded granules, and only 7 μm/day for acidogenic flocs-seeded granules. These results seem to suggest that syntrophs and *Methanothrix* species would play an important role in the formation of anaerobic granules. In fact, microbial species would differ in their capacity for aggregation, and some species are more competent for aggregation, but some are less under the same operation conditions. It seems certain that anaerobic granulation process can be expedited simply by manipulating the composition of seed sludge. This approach would be very attractive and beneficial to full-scale UASB reactor start-up. However, there is still lack of detail guidelines on which species in seed sludge should be a major component for anaerobic granulation and how to manipulate the species in seed sludge.

Characteristics of Substrate

Characteristics of feed substrate have been considered a key factor influencing the formation, composition, and structure of anaerobic granules. Based on the free energy of oxidation of organic substrate, the substrate can be roughly classified into two categories: high-energy and low-energy feeds. During the UASB start-up period, high-energy carbohydrate feeding can sustain the acidogens and facilitate the formation of extracellular polymers (Liu et al., 2002). Thus, the rapid growth of acidogens due to the presence of high-energy substrate in the influent would facilitate the overall process of sludge granulation in the UASB reactors.

Studies on mesophilic granule formation have shown that varied granular structures may be cultivated on different wastewaters and under different start-up conditions. Filamentous type granules, developed on mainly volatile fatty acid (VFA) feeds tend to be 5 mm in size and mechanically fragile. Those granules contain inert carrier material and are dominated by a highly filamentous form of *Methanothrix*, presumed to be *M. soehngenii*. More robust rod-type granules developed on sugar beet or potato processing wastewaters, and they contain no detectable inert carrier and are again dominated by *M. soehngenii*-like species, but in a much shorter chain-length of up to 3 mm in size (Adebowale and Kiff, 1988). The granules grown on VFA mixture (acetate, propionate, and butyrate) under mesophilic conditions can be classified into three distinct types according to the predominant acetate utilizing methanogens (Hulshoff Pol et al., 1983; de Zeeuw, 1984; Lettinga et al., 1984): (a) rod-type granules, which are mainly composed of rod-shaped bacteria in fragments of about 4 to 5 cells resembling *Methanothrix*; (b) filament-type granules, which consist predominantly of long multicellular rod-shaped bacteria; and (c) sarcina type granules, which develop when a high concentration of acetic acid is maintained in the reactor.

Successful formation of very small thermophilic granules (0.2 mm) from a mixture of acetic, propionic, and lactic acids had been reported (Endo and Tohay, 1988), while larger aggregates of 3.0 mm in diameter were obtained by Bochem et al. (1982) in chemostat studies of acetate enrichments. Those granules consisted of densely packed *Methanosarcina* clusters surrounding a more loosely packed central area, which contained at least two non-methanogenic species.

A trend was observed towards a wider diversity of methanogenic sub-populations paralleling an increase in the complexity of waste composition. At least four distinct micro-colonies were observed in granules treating brewery wastewater (Wu, 1991). One of these micro-colonies was composed of *Methanothrix*-like rods only, while the other micro-colonies consisted of H_2–CO_2 utilizing *Methanobacterium*-like rods in juxtaposition with three different rod-shaped syntrophs (Hickey, 1991). Based on full-scale UASB experiences in treating a variety of different wastewaters, it has been established that granulation of anaerobic sludge takes place in many different types of wastewaters. With a substrate containing 10% sucrose and 90% VFA mixture (acetate plus propionate), granular and flocculent sludge cannot be effectively separated. The granules contained a high fraction of filamentous organisms that were mainly attached to inert support particles. A feed change from a VFA mixture to a carbohydrate solution may lead to problems of flotation and formation of a rather voluminous type of sludge if the granules are cultivated on acidified wastewaters.

Chen and Lun (1993) cultivated three types of anaerobic granules with acetic acid, glucose, and alcoholic stillage, respectively, and found that the properties of three types of granules were significantly different. The anaerobic granules fed with alcoholic stillage have the better physical properties in terms of density, SVI, and intensity. This is probably due to the complexity of the substrate constituents, which leads to an abundant microbial diversity in the granules. It must be realized that the energy containing in the substrate is important for anaerobic granulation, however the complexity of substrate would exert a selection pressure on microbial diversity in anaerobic granules as discussed earlier. Such a selection pressure would influence the formation and microstructure of granules.

Chemical Conditions

Ionic composition and presence of polymer in the anaerobic system are believed to have significant roles in the forming granules through various mechanisms and models. The effects of various cations and polymers are discussed in the following sections.

Effect of Cations

Divalent and trivalent cations have positive effects on flocculation of dispersed sludge. Commonly used divalent cations are calcium and magnesium while iron can be used as both a divalent or trivalent cation depending on its oxidation state. Evidence shows that the presence of divalent and trivalent cations, such as Ca^{2+}, Mg^{2+}, Fe^{2+}, and Fe^{3+}, helps bind negatively charged cells together to form microbial nuclei that promote further granulation (Mahoney et al., 1987; Schmidt and Ahring, 1993; Teo et al., 2000; Yu et al., 2001).

The use of divalent or trivalent cation to assist in granulation lies in their ability to condense the diffusive double-layers resulting in relatively stronger effect of van der Waals attractive forces. Calcium was also found to form calcium bridge between its ion and extracellular polymers (ECP) (Forester and Lewin, 1972; Rudd et al., 1984).

According to McCarty et al. (1986), calcium stimulates granulation at concentration of 100–200 mg/l and becomes inhibitory at >2500 mg/l. Similarly, de Zeeuw (1984); Mahoney et al. (1987) reported that the rate of sludge granulation was significantly enhanced in a calcium concentration range of 100–200 mg/l. Verrier and Albagnac (1985) suggested the possibility that divalent calcium indirectly promote bacterial adhesion by increasing surface hydrophobicity.

Grotenhuis et al. (1988) found that contacting granular sludge with calcium chelating agent (EGTA) will result in granules disintegrating and becoming weaker. Based on this observation, it was then concluded that calcium plays an important role in granulation in 2 ways:

1. Inorganic calcium precipitates serve as surface for adhesion of anaerobic bacteria;
2. Calcium may be a constituent of extracellular polysaccharides and/or proteins that are not present as sticking material.

Research by Teo et al. (2000) showed that an increase in Ca^{2+} concentration from 0 to 80 mg/l substantially improved the strength of anaerobic granules, as indicated by a 60% decrease in turbidity. A study by Batstone and Keller (2001) using granules from UASB reactor was conducted to investigate the influence of calcium on granular sludge in a full scale UASB treating paper mill wastewater. It was found that the granules were small (1.0 mm) with a narrow size distribution. The core of the granules which was 200–400 micron in diameter consisted mainly of calcium precipitates. The rest of the granules were biologically active. With the observation that the core varied in consistency rather than size, it was concluded that it may have formed in the bulk liquid as amorphous calcium carbonate and subsequently acted as a nucleus for granule formation. As the granule increased in size, the calcium probably continued to precipitate in the core until saturation, after which scaling and granule deactivation occurred. At high calcium concentrations, problems such as the precipitation of calcium on the surface of granules and accumulation of calcium inside anaerobic granules with consequent reduced microbial activity have also been reported (Yu et al., 2001).

The role of cations in anaerobic granulation processes is still uncertain. Despite the positive effects reported, there were studies indicating that calcium ions did not induce sludge granulation at all (Guiot et al., 1988) and high concentration of magnesium ion (used as a divalent cation) caused granules to fall apart (Schmidt and Ahring, 1993). This may be due to the notion that, at high cation concentrations, bacteria could change their surface charge from negative to positive resulting in repulsion which deters the granulation process.

Effect of Polymers

It is generally accepted that ECP plays an important role in the formation of a supporting matrix for the microorganisms. Production of ECP is believed to be affected by the nutritional balance and/or the diversity of the granules microflora.

According to Dolfing (1985), ECP contributes to about 1–2% on a dry weight basis. Ross (1984) found that ECP accumulation plays an important role in the "clumping" of bacteria that is comparable to the role of microbiological agglutination in the flocculation of aerobic sludge. Harada et al. (1988) found that biopolymer production on acetate is limited and therefore ECP is not a prerequisite for granulation. de Zeeuw (1984) however observed high growth yield factors in batch fed reactors and UASB reactors using acetate as a single substrate. He explained this by presuming that most of the growth took place in the form of ECP production. This conclusion was supported by the observation that extra ammonia fixation could not be detected.

Synthetic and natural polymers have been widely used in coagulation/flocculation processes. These polymers are known to promote particle agglomeration and have been used to enhance the formation of anaerobic granules. The influence of synthetic polymers (Percol 763) and natural chitosan polymers on the granulation rate of suspended anaerobic sludge was studied in laboratory-scale UASB reactor (El-Mamouni et al., 1998). Results showed that reactor supplemented with either natural or synthetic polymers achieved better granulation. A greater granulation was obtained with chitosan compared to Precol 763. The superior granulation performance of chitosan may be related to its polysaccharidic structure which is similar to ECP that helps in aggregating anaerobic granules. The polymer enhanced granules had about the same specific activity of methane production as the granules formed without the polymer. Polymeric chains enhance flocculation by bridging microbial cells. Such initial microbial nuclei are the first step in microbial granulation. In short, the results showed that polymers play a critical role in enhancing anaerobic granulation in UASB-like reactors.

Kalogo et al. (2001) used a water extract of *Moringa oleifera* seeds (WEMOS) to assist in the start-up of UASB reactor. The ability of WEMOS to adsorb on the surface of dispersed bacteria which eventually lead to neutralization of their surface charge assist in the granulation process.

In cationic polymer-assisted anaerobic granulation processes, it has been observed that the start-up period required for the development of granular sludge blanket can be shortened significantly compared to when no polymers are used (Uyanik et al., 2002). Two mechanisms appear to be involved in polymer enhancement of anaerobic granulation. The addition of polymers to anaerobic systems likely changes the surface properties of bacteria to promote association of individual cells. Polymer may also form a relatively solid and stable three-dimensional matrix within which bacteria multiply and daughter cells are then confined. The polymer additives appear to play a similar role as do the naturally secreted extracellular polymeric substances (EPS) in aggregating anaerobic sludge.

Show et al. (2004) investigated the influence of a coagulant polymer on start-up, sludge granulation, and the associated reactor performance in laboratory-scale UASB reactors. The experimental results demonstrated that adding the polymer at an appropriate dosage markedly accelerated the start-up time. The time required to reach stable treatment at an organic loading rate (OLR) of 4 g COD/l.d was reduced by approximately 50% as compared with the control reactor, while other reactors also recorded varying degree of shortening. Monitoring on granule development showed that the granule formation was accelerated by 30% from the use of the appropriate dosage of polymer. Subsequent granules characterization indicated that granules developed in the polymer-assisted reactor exhibited the best settleability, strength, and methanogenic activity at all OLRs. The organic loading capacities of reactors were also increased by the polymer addition to as high as 40 g COD/l.d. The laboratory results obtained demonstrated that adding the cationic polymer could result in shortening of start-up time and enhancement of granulation, which in turn lead to improvement in organics removal efficiency and loading capacity of the UASB system. The authors hypothesized that positively charged polymer form bridges among the negatively charged bacterial cells through electrostatic charge attraction. The bridging effect would enable greater interaction between biosolids resulting in preferential development and enhancement of biogranulation in UASB reactors.

Summary of Recommendations for Developing Granular Sludge

Biological Aspects

1. The seed sludge for start-up should be granulated and acclimatized for the type of waste to be treated.
2. If (1) is not possible, then the seed sludge must contain as much variety of microorganisms as possible.
3. Biological loading rate during start up should not be excessive (<0.6 kg COD/kg VSS per day).
4. The presence of any kind of toxic compounds should be avoided. If toxic waste is to be treated, it should be acclimatized.

Chemical Aspects

1. Divalent salts should be added as this has been proven to be beneficial to some extent.
2. Granulation will occur faster on mainly soluble unacidified wastewaters than on acidified wastewaters. The pH should be kept as near to neutrality as possible even though granulation has been reported to occur at pH 5–6.
3. Sodium carbonate and phosphorus can be used to maintain alkalinity and buffer conditions, respectively in the reactor.
4. All essential growth factors such as N, P, S, and trace elements (Fe, Ni, Co) should be present in sufficient amounts and in available form.

Physical Aspects

1. Main physical aspects are related to reactor's design and temperature. Design can directly or indirectly affect the mixing characteristics of the reactor, the liquid and gas upflow velocity rates, and even the shape of the biomass.

2. Temperature, besides affecting the physio-chemical fact described above, also affects the chemical–biological reaction rates. The optimal temperature for mesophilic treatment is in the range of 30–38°C and 50–60°C for thermophilic treatment.

Wastewater Characteristics

1. Lower strength of wastewater can be used. It was observed the lower the strength of the wastewater, the faster granulation will proceed. The strength should however be high enough to maintain good conditions for bacterial growth. The minimum COD level is presumably approximately 500 mg/l.
2. Dispersed solids such as acidogenic biomass and fibrous matter, retard or may even prevent granulation. Hence the presence of the above should be avoided or minimized.

References

Adebowale, O., & Kiff, R. (1988). Operational trends in UASB reactor bed stability and during initiation of granulation (eds. Tilche, A., & Rozzi, A.), *Proceeding of the Fifth International Symposium on Anaerobic Digestion*, Bologna, Italy, 99–103.

Alphenaar, P.A., Visser, A., & Lettinga, G. (1993). The effect of liquid upflow velocity and hydraulic retention time on granulation in UASB reactors treating wastewater with a high-sulphate content. *Bioresour. Technol.*, *43*, 249–258.

Alves, M., Cavaleiro, A.J., Ferreira, E.C., Amaral, A.L., Mota, M., da Motta, M., Vivier, H., & Pons, M.N. (2000). Characterization by image analysis of anaerobic sludge under shock conditions. *Water Sci. Technol.*, *41*, 207–214.

Angenent, L.T., Banik, G.C., & Sung, S.W. (2001). Anaerobic migrating blanket reactor treatment of low-strength wastewater at low temperatures. *Water Environ. Res.*, *73*, 567–574.

Arcand, Y., Guitot, S.R., Desrochers, M., & Chavarie, C. (1994). Impact of the reactor hydrodynamics and organic loading on the size and activity of anaerobic granules. *Chem. Eng. J. Biochem. Eng. J.*, *56*, 23–35.

Batsone, D.J., & Keller, J. (2001). Variation of bulk properties of anaerobic granules with wastewater type. *Water Res.*, *35*, 1723–1729.

Bitton, G., (1999). *Wastewater Microbiology*. Wiley–Liss, New York.

Bochem, H.P., Schoberth, S.M., Sprey, B., & Wengher, P. (1982). Thermophilic biomethanation of acetic acid: morphology and ultrastructure of granular consortium. *Can. J. Microbiol.*, *28*, 500–510.

Campos, C.M.M., & Anderson, G.K. (1992). The effect of the liquid upflow velocity and the substrate concentration on the startup and steady state period of lab-scale UASB reactors. *Water Sci. Technol.*, *25*, 41–50.

Chen, J., & Lun, S.Y. (1993). Study on mechanism of anaerobic sludge granulation in UASB reactors. *Water Sci. Technol.*, *28*, 171–178.

de Zeeuw, W.J. (1984). Acclimatization of anaerobic sludge for UASB reactor start-up, *Ph.D. Thesis*. Agricultural University of Wageningen, The Netherlands.

de Zeeuw, W.J. (1988). Granular sludge in UASB-reactors. *Granular Anaerobic Sludge: Microbiology and Technology* (eds. Lettinga, G., Zehnder, A.J.B., Grotenhuis, J.T.C., & Hulshoff Pol, L.W.), Wageningen, The Netherlands, 132–145.

Dolfing, J. (1985). Kinetics of methane formation by granular sludge at low substrate concentrations: The influence of mass transfer limitation. *Appl. Microbiol. Biotechnol.*, *22*, 77–81.

El-Mamouni, R., Leduc, R., & Guiot, S.R. (1997). Influence of starting microbial nucleus type on the anaerobic granulation dynamics. *Appl. Microbiol. Technol.*, *47*, 189–194.

El-Mamouni, R., Leduc, R., & Guiot, S.R. (1998). Influence of synthetic and natural polymers on the anaerobic granulation process. *Water Sci. Technol.*, *38*, 341–347.

Endo, G., & Tohay, Y. (1988). Ecological study on anaerobic sludge bulking caused by filamentous bacteria growth in an anaerobic contact process. *Water Sci. Technol.*, *20*, 205–211.

Fang, H.H.P., & Chui, H.K. (1993). Maximum COD loading capacity in UASB reactors at 37°C. *J. Environ. Eng.*, *119*, 103–119.

Fang, H.H.P., & Lau, I.W.C. (1996). Start up of thermophilic UASB reactors using different mesophilic seed sludges. *Water Sci. Technol.*, *34*, 445–452.

Forester, C.F., & Lewin, D.C. (1972). Polymer interaction at activated sludge surfaces. *Effluent Water Treat. J.*, *12*, 520–525.

Ghangrekar, M.M., Asolekar, S.R., Ranganathan, K.R., & Joshi, S.G. (1996). Experience with UASB reactor start-up under different operation conditions. *Water Sci. Technol.*, *34*, 421–428.

Grotenhuis, J.T.C., van Lier J.B., Plugge C.M., & Stams A.J.M. (1988). Effect of calcium removal on size and strength of methanogenic granules. (ed. Monduzzi), *Poster-papers of 5th International Symposium on Anaerobic Digestion*, Bologna, Italy, 117–120.

Guiot, S.R., Gorur, S.S., & Kennedy, K.J. (1988). Nutritional and environmental factors contributing to microbial aggregation during UFB reactor start-up. (eds. Hall, E.R., & Hobson, P.N.), *Anaerobic Digestion*, Pergamon Press 47–53.

Harada, H., Endo, G., Tohya, Y., & Momonoi, K. (1988). High rate performance and its related characteristics of granulated sludges in UASB reactors treating various wastewater. *Proceedings of 5th International Symposium on Anaerobic Digestion*. Bologna, Italy, May, 22–26, 1986.

Hickey, R.F. (1991). Startup, operation, monitoring, and control of high-rate anaerobic treatment systems. *Water Sci. Technol., 24*, 207–255.

Hulshoff Pol, L.W. (1989). The phenomenon of granulation of anaerobic sludge, *Ph.D. Thesis*. Agricultural University of Wageningen, The Netherlands.

Hulshoff Pol, L.W., de Zeeuw, W.J., Velzebber, C.T.M., & Lettinga, G. (1983). Granulation in UASB-reactors. *Water Sci. Technol., 15* (8/9), 291–304.

Hulshoff Pol, L.W., Heijnekamp, K., & Lettinga, G. (1988). The selection pressure as a driving force behind the granulation of anaerobic sludge. *Granular Anaerobic Sludge: Microbiology and Technology* (eds. Lettinga, G., Zehnder, A.J.B., Grotenhuis, J.T.C., & Hulshoff Pol, L.W.) GASMAT, Lunteren, Wageningen, 153–161.

Kalogo, Y., Seka, A.M., & Veretraete, W. (2001). Enhancing the start-up of a UASB treating domestic wastewater by adding a water extract of *Moringa oleifera* seeds. *Appl. Microbiol. and Biotechnol., 55*, 644–651.

Kosaric, N., Blaszczyk, R., Orphan, L., & Valladares, J. (1990). The characteristics of granules from upflow anaerobic sludge blanket reactors. *Water Res., 24*, 1473–1477.

Lepisto, R., & Rintala, J. (1999). Extreme thermophilic (70°C), VFA-fed UASB reactor: performance, temperature response, load potential, and comparison with 35 and 55°C UASB reactors. *Water Res., 33*, 3162–3170.

Lettinga, G., Hulshoff Pol, L.W., Koster, I.W., Wiegant, W.M., Zeeuw, W.J., de Rinzema, A., Grin, P.C., Roersma, R.E., & Hobma, S.W. (1984). High-rate anaerobic wastewater treatment using the UASB reactor under a wide range of temperature conditions. *Biotechnol. Genet. Eng. Rev., 2*, 253–284.

Lettinga, G., Rebac, S., & Zeeman, G. (2001). Challenge of psychrophilic anaerobic wastewater treatment. *Trends Biotechnol., 19*, 363–370.

Lew, B., Belavski, M., Admon, S., Tarre, S., & Green, M. (2003). Temperature effect on UASB reactor operation for domestic wastewater treatment in temperate climate regions. *Water Sci. Technol., 48*, 25–30.

Liu, Y., & Tay, J.H. (2001). Detachment forces and their influence on the structure and metabolic behavior of biofilms. *World J. Microbiol. Biotechnol., 17*, 111–117.

Liu, Y., Xu, H.L., Show, K.Y., & Tay, J.H. (2002). Anaerobic granulation technology for wastewater treatment. *World J. Microbiol. Biotechnol.*, *18* (2), 99–113.

Mahoney, E.M., Varangu, L.K., Cairns, W.L., Kosaric, N., & Murray, R.G.E. (1987). Effect of Ca^{2+} on microbial aggregation during UASB reactor start-up. *Water Sci. Technol.*, *19*, 249–260.

McCarty, P.L., & Smith, D.P. (1986). Anaerobic wastewater treatment. *Env. Sci. Tech.*, *20* (12), 1200–1206.

Morvai, L., Mihaltz, P., & Czako, L. (1992). The kinetic basis of a new start-up method to ensure the rapid granulation of anaerobic sludge. *Water Sci. Technol.*, *25*, 113–122.

Noyola, A., & Mereno, G. (1994). Granulation production from raw waste activated sludge. *Water Sci. Technol.*, *30*, 339–346.

O'Flaherty, V., Lens, P.N., de Beer, D., & Colleran, E. (1997). Effect of feed composition and upflow velocity on aggregate characteristics in anaerobic upflow reactors. *Appl. Microbiol. Biotechnol.*, *47*, 102–107.

Qi, S., Dhavises, G., Sriprasetsak, P., Tanaka, T., & Makoto, T. (1985). Methane fermentation of agro-waste and grasses. *In Anaerobic Digestion 1985*, (ed. China State Biogas Association), Guangzhou, China, 145–157.

Quarmby, J., & Forester, C.F. (1995). An examination of the structure of UASB granules. *Water Res.*, *29*, 2449–2454.

Ross, W.R. (1984). The phenomenon of sludge pelletization in the treatment of maize processing waste. *Water SA*, *10*, 197–204.

Rudd, T., Sterritt, R.M., & Lester, J.N. (1984). Complexation of heavy metals by extracellular polymers in activated sludge process. *J. Water Pollut. Control Fed.*, *56* (99), 1260–1268.

Schmidt, J.E., & Ahring, B.K. (1993). Effects of magnesium on thermophilic acetate-degrading granules in upflow anaerobic sludge blanket (UASB) reactor. *Enzyme Microb. Technol.*, *15*, 304–310.

Show, K.Y., Tay, J.H., Yang, L., Wang, Y., & Lua, C.H. (2004). Effects of stressed loading on start-up and granulation in UASB reactors. *J. Environ. Eng.*, *130* (7), 743–750.

Sierra-Alvarez, R., Hulshoff Pol, L.W., & Lettinga, G. (1988). Start-up of a UASB reactor on a carbohydrate substrate. *Granular Anaerobic Sludge: Microbiology and Technology* (eds. Lettinga, G., Zehnder, A.J.B., Grotenhuis, J.T.C., & Hulshoff Pol, L.W.), Wageningen, The Netherlands, 223–227.

Singh, K.S., & Viraraghavan, T. (2003). Impact of temperature on performance, microbiological, and hydrodynamic aspects of UASB reactors treating municipal wastewater. *Water Sci. Technol.*, *48*, 211–217.

Tay, J.H., & Yan, Y.G. (1996). Influence of substrate concentration on microbial selection and granulation during start-up of upflow anaerobic sludge blanket reactors. *Water Environ. Res.*, *68*, 1140–1150.

Tay, J.H., Xu, H.L., & Teo, K.C. (2000). Molecular mechanism of granulation: I. H+ translocation–dehydration theory. *J. Environ. Eng.*, *126*, 403–410.

Teo, K.C., Xu, H.L., & Tay, J.H. (2000). Microscopic observation of aerobic granulation in sequential aerobic sludge reactor. *J. Appl. Microbiol.*, *91*, 168–175.

Uyanik, S., Sallis, P.J., & Anderson, G.K. (2002). The effect of polymer addition on granulation in an anaerobic baffled reactor (ABR): Part I. Process performance. *Water Res.*, *36*, 933–942.

Verrier, D., & Albagnac, G. (1985). Adhesion of anaerobic bacteria from methanogenic sludge onto inert solid surfaces. *Paper presented at the EEC-Conference Energy from Biomass*, Venice, Italy, 25–29 March.

Wiegant, W.M. (1986). Thermophilic anaerobic digestion for waste and wastewater treatment. *Ph.D. Thesis*, Agricultural University of Wageningen, Wageningen, The Netherlands.

Wu, W.M. (1991). Technological and microbiological aspects of anaerobic granules. *Ph.D. Dissertation*, Michigan State University, MI, U.S.

Wu, W.M., Hu, J.C., & Gu, X.S. (1985). Properties of granular sludge in upflow anaerobic sludge blanket (UASB) reactor and its formation. *Anaerobic Digestion 1985* (ed. China State Biogas Association), Guangzhou, China, 339–354.

Wu, W.M., Hu, J.C., Gu, X.S., & Gu, G.W. (1987). Cultivation of anaerobic granular sludge in UASB reactors with aerobic activated sludge as seed. *Water Res.*, *21*, 789–799.

Xu, H.L., & Tay, J.H. (2002). Anaerobic granulation with methanol-cultured seed sludge. *J. Environ. Sci. Health, Part A*, *37*, 85–94.

Yu, H.Q., Tay, J.H., & Fang, H.H.P. (2001). The role of calcium in sludge granulation during UASB reactor start-up. *Water Res.*, *35*, 1052–1060.

Zeikus, J.G. (1979). Microbial populations in digesters. *In Anaerobic Digestion* (ed. Stafford, A.D.), Scientific Press, Cardiff, 75–103.

Chapter 3

Applications of Anaerobic Granulation

Kuan-Yeow Show

Introduction

Major contributions to the broad application of anaerobic granulation treatment and a better understanding of this process are a result of much effort of the researchers in the 1980s. Much emphasis has been placed on the significance of anaerobic granulation technology in meeting the need for sustainable development in the future. Substantial effort is now being installed in exploring broader applications of anaerobic granulation treatment for removal of unwanted organic pollutants by converting them into biogas, a renewable energy source.

Anaerobic treatment using sludge granulation has gained tremendous success over the past two decades for treatment of a variety of industrial effluents. Apparent advantages derived are from low operating costs, compact reactor construction, production of energy in the form of biogas, low surplus sludge production, which result in overall favorable economics. For anaerobic treatment to compete with alternative technologies such as aerobic or physico-chemical treatment, it has to be cost-effective in terms of investment and operating costs, reliable, and durable. System designs have focused on increased process control to secure optimal operating conditions and system "compactness" in order to reduce investment costs.

High-rate anaerobic process for industrial wastewater was first applied on a commercial-scale in the sugar industry in the mid-1970s. Since then, the technology has developed into a well-received method of wastewater treatment for a wide variety of industries. The technology is now employed in over 65 countries and a total of approximately 1400 plants were built by the 16 leading vendors of such systems (Frankin, 2001). These high-rate treatment plants account for approximately 65% of the total number of anaerobic treatment plants for industrial applications, which is estimated to be 2000. From an analysis of a database consisting of 1215 plants, it appears that the upflow anaerobic sludge blanket (UASB) technology as originally developed in the Netherlands is the most predominant process. It is also seen that the highly loaded expanded granular sludge bed systems are gradually replacing at least some of the upflow anaerobic sludge blanket applications.

Various reactor designs have been developed over the past two decades that are based on various ways of retaining biomass within the reactor system. Recently, information on implementation of anaerobic technologies used for the treatment of municipal and industrial wastes and wastewaters was collected (Hulshoff Pol et al., 1998). A comprehensive overview is given of the anaerobic systems and advancement currently used for the worldwide treatment industrial wastewaters.

Types of Anaerobic Treatment Plants Installed Worldwide

The key to successful application of anaerobic treatment is to un-couple the hydraulic retention time (of wastewater) and the solids retention time (of active biomass) in the reactor system. In order to achieve high system loading rates, short hydraulic retention times should be applied, whilst at the same time maintaining positive net solids (biomass) retention. Thus, various reactor designs were developed over the past two decades that are based on various ways of retaining biomass within the reactor system. Table 3.1 shows the number of plants categorized for different anaerobic processes.

Scope of Applications

Anaerobic treatment process has been successfully applied in various types of industrial wastewaters. It ranges from food wastewaters to non-food

Table 3.1. Processes used by vendors in database (Frankin, 2001)

Process	Number of plants	% average in database	% average over 1990–1996	% average over 1997–2000
UASB	682	58	68	34
EGSB	198	16	8	50
Low rate† (lagoon/contact)	187	15	12	8
Fixed bed†	54	4	4	3
Fluidized bed†	16	1	2	1
Hybrid	12	1	1	2
Undefined/unknown	66	5	6	3
Total in database	1215	100	101	101

†Denotes non-granular type of anaerobic system.

Table 3.2. Applications served by vendors in database (Frankin, 2001)

Application	Number of plants	%
Breweries and beverages	329	27
Distilleries and fermentation	208	17
Chemical	63	5
Pulp and paper	130	11
Food	389	32
Landfill leachate	20	2
Undefined/unknown	76	6
Total in database	1215	100

wastewater, such as chemical plants. Table 3.2 illustrates that food, breweries and beverages industries have the largest share of anaerobic treatment plants built to treat their wastewaters. The data presented also affirms that anaerobic treatment is an established technology for a wide variety of industrial applications.

The fact that chemical wastewater can be treated using anaerobic process is very important and encouraging. Recent advances in chemical technology have led to the production of many and new potentially dangerous compounds called xenobiotics in the wastewater streams. These new substances, representing a wide array of compounds ranging from phenols to pesticides, each presents its own individual problems when dealing with its ultimate disposal. Many new and innovative techniques for the treatment of these wastes have been introduced, many of which employ anaerobic processes.

Brewery, beverages, distillations, and fermentation wastewaters present problems similar to those of sugar processing, namely high organic strength. These wastewaters are composed of the residues of the fermented material which contributes to high biological oxygen demand (BOD), as high as 25,000 mg/L after the alcohol has been distilled off. Anaerobic processes have been used to treat these wastewaters, with good results and thus this explains the large percentage (44%) of anaerobic application in this area.

Food processing industries cover a diverse scope from fish processing, pear and pineapple wastes to bean blanching wastes. Anaerobic treatment process has proven to yield encouraging results. Again, almost one-third of the anaerobic process plants are dedicated to treat the food wastewaters.

Applications of Anaerobic Granulation

Upflow Anaerobic Sludge Blanket Reactor (UASB)

The sludge blanket concept was first used in the Reversed Flow Dorr Oliver clarigester which is a modified version of the contact process. Unlike the contact process, it is unmixed and feed flows upward through a dense bed of flocculated bacteria. Flocs collect in the settling compartment and return to the reactor by gravity. Dorr Oliver clarigester improvements due to the biomass loss in effluent led to the upflow sludge blanket process and the UASB process. The UASB has an integrated 3-phase liquid–gas–solids separator to help retain sludge, and mechanical agitation is minimized or omitted.

The UASB reactors have been applied to a wide range of industrial wastewater, including those containing toxic or inhibitory compounds. The process is also feasible to treat domestic wastewater with temperature as low as 14–16°C or even lower.

Since the early 1980s, considerable research and development has occurred in relation to anaerobic wastewater treatment systems and specifically, UASB. Reductions in BOD of 75–90% have been noted in tropical conditions. The UASB technology is feasible in an urban, developing world context because of its high organic removal efficiency, simplicity, low cost, low capital and maintenance costs, and low land requirements. Anaerobic treatment processes are suitable in tropical conditions because anaerobic treatment functions well in temperatures exceeding 20°C.

They are characterized by low sludge production and low energy needs. The UASB is typically constructed with entrance pipes delivering influent to the bottom of the unit and a 3-phase separator at the top of the reactor to separate the biogas from the liquid phase (water and sludge) and of sludge from the water phase; overall this prevents sludge washout.

The UASB process was first described in an international journal with studies on the treatment of methanol with laboratory UASB reactors (Lettinga et al., 1979a), and its potential for dilute wastewaters in general was described (Lettinga et al., 1979b, 1980). There emerged subsequently many reports by Lettinga and his colleagues on application of the UASB process for treatment of a variety of wastewaters including those from sugarbeets (Lettinga et al., 1976, 1977), piggery (van Velsen et al., 1979), alcohols (Lettinga et al., 1979a), fatty acids (Ten Brummeler et al., 1985, Hwu et al., 1997a,b,c), slaughterhouse (Sayed et al., 1987, 1988), potato starch (Field et al., 1987), milk fat (Petruy and Lettinga, 1997; Petruy et al., 1997), and pulp and paper wastes (Sierra-Alvarez et al., 1990; Lettinga et al., 1991; Rintala et al., 1991). A comparison of performance of several full-scale UASB reactors is summarized in Table 3.3.

Of growing interest is the application of the UASB process for treatment of domestic wastewaters, which they clearly demonstrated is feasible (Lettinga et al., 1983, 1993; van der Last and Lettinga, 1992; Bogte et al., 1993; Lettinga, 1995).

The following section highlights the most recent full-scale and pilot-scale findings in the anaerobic treatment of industrial wastewaters.

A control system that is able to manage the start-up of a UASB reactor, using a reduced number of process variables was developed (Punal et al., 2001). Two different start-up strategies were applied: fed-batch and continuous operation. In the fed-batch, results show that starting from an organic loading rate (OLR) of lower than 0.5 kg COD/m^3·d, a load of higher than 8 kg COD/m^3·d was achieved in only 33 days and the COD removal efficiency was over 90%. In the continuous system, results showed 24 h of an excellent value, and also, starting from an OLR of

Table 3.3. The performance of full-scale UASB reactors (extracted from Liu et al., 2002)

Reactor volume (m^3)	Wastewater type	Influent COD (g/l)	COD removal (%)	BOD removal (%)	Biogas production	References
1500	Juice	22	88	96	0.5 m^3/kg COD	Driessen et al. (1994)
1500	Molasses	80–120	65–70	85–90	0.5 m^3/kg COD	Driessen et al. (1994)
120	Sewage	0.4	74	80		Vieira et al. (1994)
1400	Brewery	1.7	80		2.0 m^3/m^3 d	Pereboom and Vereilken (1994)
2200	Paper mill	1.3	70		1.4 m^3/m^3 d	Pereboom and Vereilken (1994)
1700 × 2 unit	Potato processing	12	95		3.0 m^3/m^3 d	Pereboom and Vereilken (1994)
1200	Sewage	0.56	68–74	69–76	0.05–0.1 m^3/kg COD	Draaijer et al. (1992)
120	Sewage	0.19–0.46	55–65	60–72	0.09–0.25 m^3/kg COD	Vieira and Garcia (1992)

lower than 0.5 kg COD/m^3·d, a load of 9–12 kg COD/m^3·d was achieved in 40 days and the COD removal efficiency was over 95%. Comparing the standard deviation of the process parameters, fed-batch mode has a better process efficiency. However, continuous mode has a better capacity to treat the organic load by enabling the system to operate at a more stable influent OLR. This is especially useful during the first 2 weeks of the start-up phase.

Another way to have a fast start-up is to adjust the microbial load index (MLI) values (Tay and Yan, 1997). Findings show that under high MLIs of 0.8 and 0.6, granulation developed well in 3 to 4 months of operation, making a fast start-up. However, with low MLIs of 0.3 and 0.2, there was still no granulation after 6 months. It was observed that there are 3 phases during the process of granulation, namely acclimation, granulation, and maturation. A stepwise and gradual increment in the sludge loading rate (SLR) must be followed too, to avoid the scenario of overloading or starving at different stages.

Show et al. (2004a) had developed an unconventional approach to accelerate start-up and granulation processes in UASB reactors by stressing the organic loading rate (OLR) without having to reach steady-state conditions. Three UASB reactors treating a synthetic feed with chemical oxygen demand (COD) of 2500 mg/L, at a mesophilic temperature of 35°C were studied. The results indicate that the start-up of reactors could be significantly accelerated under stressed loading conditions. Start-up times of the moderately and severely stressed reactors for operating at OLRs of 1 to 16 g COD/L·d ranged from 10 to 80 days and 13 to 90 days, respectively. Comparing with 17 to 120 days needed in the control reactor to reach the same OLRs, the start-up times were shortened by 25 to 41%. The extent of acceleration depends on the level at which the reactor are stressed. Applying stress and the extent of stress level in starting up the reactors did not reduce the reactor loading capacity, as all the reactors reached a similar maximum OLR of 16 g/L·d at the end of operation.

Show et al. (2004a) explored further into the possibility of accelerating development of granulation with the unconventional approach of stressed loadings. The researchers found that under stressed loading conditions, the sludge particles began to form granules earlier in both the stressed reactors after 24 and 30 days of start-up operation. Comparing with the control reactor without applying stress, the time taken to form granule was reduced by 45 and 32% in the severely and moderately stressed units, respectively. The granule formation occurred earlier in the severely stressed reactor than the moderately stressed unit.

While the results obtained had established significant acceleration in start-up and granulation processes, the characteriztics of granules developed were greatly influenced by the level of stress exerted. Characterization of bioparticles revealed that the granules developed in the moderately stressed reactor exhibited superior characteristics in terms of settleability, strength, microbial activity and morphology, and granular sludge growth, as compared with both the control reactor operated without stress and the unit which was over-stressed (Show et al., 2004a). The results presented indicate that the unconventional start-up approach could offer a practical solution for the inherent long start-up in UASB systems with concomitant saving in time and cost.

The microbial mechanism of thermophilic granulation and sludge retention during start-up was studied (Syutsubo et al., 1998). Development of well-settleable granular sludge is the key factor of successful operation of the UASB process. The inoculum was taken from thermophilic digested sewage sludge. Operating temperature was 55°C. The feed solution consisted of sucrose, yeast, and volatile fatty acids which are acetate and propionate. Results showed that the granule's sludge volume index (SVI) finally settled at about 13 mL/g volatile suspended solid (VSS) upon maturation of the thermophilic granules. As a result of establishment of the entire granulated sludge bed, the reactor allowed a maximum volumetric COD loading of 45 kg COD/m^3·d with a COD removal efficiency of 90%. The maximum sludge loading achieved was 3.7 g COD/g VSS·d, which was two to three times larger than that of sludge grown under mesophilic conditions.

Both acetate and hydrogen utilizing methanogenic activities exhibited their optima at 65°C, while that of propionate-fed methanogenic activity was at 50°C. Methanogenic activities of the retained sludge increased finally up to 110 times for acetate, 25 times for propionate, and 3.6 times for hydrogen, when compared with those of the seeded sludge. This relatively low value for propionate implies that the propionate degradation was most likely to be a rate-limiting step in the thermophilic anaerobic process.

Early development of UASB in the 1980s had been used in treating food industry wastewaters, such as beet sugar, corn, and potato starch processing. Recent studies showed that UASB can be applied in treating wastewaters containing concentrated proteins (Fang, 1994) and aromatic compounds such as phenol (Wen et al., 1995). Changing the rate of effluent recirculation is widely used to avoid toxic impact to the microorganisms. Recirculation, together with biogas production, results in higher superficial upflow velocity that causes washout of biomass. Low hydraulic loading rate on treatment of wastewater containing high concentration of phenol using a Re-circulated UASB (RUASB) operating under mesophilic condition is encouraged (Lay and Cheng, 1998). As the hydraulic loading rates decreased from 2.5 to 1.6 m/h, the relative bacterial activity also decreased from 80 to 50%.

Granulation is generally a slow process, but it is the pre-requisite of the optimum performance of UASB-like reactors. Use of polymers enhanced the anaerobic granulation process (Mamouni et al., 1998). Chitosan, natural polymer, out-performed Percol 763, a synthetic polymer in terms of granules formation rate. Chitosan yielded a granulation rate as high as 56 m/d, compared to 35 m/d with Percol 763 in acidic pH. Under alkaline conditions, chitosan is progressively neutralized, thus resulting in a less effective flocculation of suspended sludge. The high granulation yield of chitosan was most probably attributed to its polysaccharides structure, acting similarly to the extracellular polymeric substances (EPS) in aggregating anaerobic sludge.

Recently, Show et al. (2004b) investigated the effects of cationic polymers on reactor start-up and granule development. The experimental results demonstrated that adding the polymer in the seeding stage markedly accelerated the start-up time by as much as 50% and the granule formation by 30% through the use of an appropriate dosage of polymer. In addition, granules developed in polymer-assisted reactor exhibited better settleability, strength, and methanogenic activity at all organic loading rates tested. Organic loading capacities of polymer-assisted reactor were also increased significantly from 24 to 40 g COD/L·d. It was hypothesized that the cationic polymer is able to form bridges among the negatively charged bacterial cells. The bridging enables greater interaction between biosolids resulting in preferential development and enhancement of biogranulation in UASB reactors.

Competition between methanogenesis and sulfidogenesis in anaerobic wastewater treatment exists (Zhou and Fang, 1998). High concentrations of sulfate in wastewater can adversely affect the methane production in anaerobic treatment processes. Sulfidogens degrade substrate into bicarbonates and intermediates in the reduction process of sulfate to sulfide. Sulfidogens and methanogens coexist in many anaerobic ecosystems as they have similar physiologies. They are strictly anaerobes, and in favor of similar optimum temperature and pH. Results showed that after acclimation, a benzoate removal efficiency of 99.5% was consistent regardless of the sulfate concentrations. Sulfidogenesis slowly out-competed methanogenesis during the acclimation phase. This was indicated by the increased sulfate reducing efficiency from 48 to 99% while it was accompanied by the decrease in methane production from 1.02 to 0.39 L methane/L·d.

Supplement glucose improves the anaerobic degradation of phenol (Tay et al., 2001). Phenol is present in the wastewater of some industries, like coal gasification, coke production, pharmaceutical, pesticide, fertilizer, dye manufacturing, synthetic chemical, and pulp and paper. The maximum concentration of phenol could go as high as 6000 mg/L and this is far too toxic to living aquatic organisms. Glucose is used as a co-substrate to achieve effective and constant anaerobic biodegradation of phenol. Phenol can be degraded to methane and carbon dioxide through phenol metabolizers and hydrogen utilizing and acetotrophic methanogens. The phenol removal efficiency was also the best at 98%, compared with 88% without glucose supplement. Moreover, it also exhibited greater resistance to those adverse conditions and the system recovered faster than the other system without the glucose supplement.

The results of the pilot study together with the results from the intensive laboratory studies suggest the feasibility of thermophilic anaerobic treatment for the food industry wastewaters (Rintala and Lepisto, 1997). The reactor was operated at 55°C and placed on the premises of a factory manufacturing deep-frozen goods from vegetables. The hot (greater than 80–90°C) and concentrated (14–79 g COD/L) wastewater streams, deriving from steam peeling and blanching of carrot and potato were used. More than 80% COD removal was reported.

Removal of chlorinated phenols (CP) is possible in UASB reactors (Droste et al., 1998). Halogenated organic pollutants are labeled as toxic and recalcitrant in the environment. Effluents containing CPs and related compounds are especially problematic to treat due to their persistence and their high solubility in fat. Once introduced into water ecosystems, accumulation within river sediments and bioaccumulation within the tissues of organism are possible. CP compounds were able to be metabolized to mineral end products to a large extent at loading rates where the reactor's performance was not hindered. There was no accumulation of phenol in any of the reactors in the experimental conditions.

Treatment of polyethylene terephthalate (PET) wastewater with UASB is proven feasible in full-scale application (Polanco et al., 1999). PET is generated by direct esterification of terephthalic acid (TPA) with ethylene glycol. The raw material of highly purified TPA is easily available in the market. So the wastewater from the esterification process consists of mainly unreacted raw material, largely ethylene glycol, and products of the secondary or degradation reactions, such as terephthalic acid esters, methanol, acetaldehyde, and crotonaldehyde being the major part. There is also another wastewater stream, called the second stream, from the melt spinning process where a bath of chemicals is showered to improve the physical characteristics of the fiber. It was reported that the anaerobic biodegradability was 90 and 75% for esterification wastewater and second stream wastewater respectively.

Anaerobic treatment of wastewater from a fish-canning factory is also proven feasible in a full-scale UASB reactor (Punal and Lema, 1999). The wastewater comes from two main streams, mussel cooking which is seasonal, and tuna cooking. Most of the organic load from mussel cooking wastewater consisted of carbohydrate (74.5%) while that of tuna cooking wastewater has a significant percentage of fat (23.5%) and protein (73.0%). So, the high fluctuation in wastewater characteristics caused high variance in the reactor's efficiency. However, the performance was better when a mixture of both streams was treated due to the high degradable carbohydrate content of the mussel cooking wastewater. Through alkalinity control, it was possible to operate the system properly with a COD removal between 70–90% for influent ranging from 2 to 8 kg COD/m^3·d.

The UASB technology can also treat crab-processing wastewaters (Boardman and McVeigh, 1997). Crab cooker wastewater contains high concentrations of COD, total suspended solids (TSS) and total Kjeldahl nitrogen (TKN). With UASB, the BOD_5 and COD removal efficiency was over 90%. Acidification of the feed wastewater improved treatment as it reduced the concentrations of the feed suspended solids.

It is feasible for UASB to treat tapioca starch industry wastewater effectively (Annachhatre and Amatya, 2000). After removal of suspended solids by simple gravity settling, starch wastewater was used as a feed. COD conversion efficiency was greater than 95% and gas productivity of

5–8 m^3 biogas/m^3·d was obtained. Removal of starch solids from wastewater by a simple gravity settling was sufficient to obtain satisfactory performance of the UASB process.

The application of UASB reactor in the world shows an increasing trend. However, the start-up and control of those anaerobic systems are complex due to the low methanogenic activity of microorganisms. Although a number of mathematical models for anaerobic processes are available in the literature (Harper and Suidan, 1991), it seems still difficult to quantitatively describe the anaerobic processes because of the biological nature of degradation mechanisms involved. Recently, two powerful mathematical tools, namely fuzzy pattern and neural network have been successfully introduced into the anaerobic systems, such as anaerobic filters, UASB, and expanded granular sludge bed reactor (EGSB) (Marsili-Libelli and Muller, 1996; Guwy et al., 1997; Tay and Zhang, 1999, 2000).

A database containing system performance information is a prerequisite for training the neural fuzzy model, and the performance of the neural fuzzy model is highly dependent on the quality of training data although the training data collection sometimes is quite difficult, especially for a novel system. On the other hand, the input and output parameters need to be carefully selected or generated from the parameters commonly used for anaerobic system description in order to compute the neural fuzzy model. Tay and Zhang (2000) suggested that for the liquid phase, information on pH, total and specific volatile fatty acids (VFA), alkalinity, COD or total organic carbon (TOC) concentration, COD or TOC reduction, and redox potential (ORP) must be provided; as for the gas phase, the parameters include gas production rates and methane, carbon dioxide, hydrogen, and monoxide concentrations. The advantages of the neural fuzzy model include high adaptability to the variation of system configurations. This mathematical methodology has a prospective industrial application potential for the simulation and real-time control of complex anaerobic systems.

Expanded Granular Sludge Bed Reactor (EGSB)

Anaerobic sludge bed (and in particular the EGSB) reactor systems can be started up within a few days with granular seed sludges, and they may be applied across a wide range of conditions and strengths of wastewater (Lettinga, 1996). EGSB systems are particularly suited to low temperatures (10°C) and very low strengths (very much smaller than 1000 mg/L) and for the treatment of recalcitrant or toxic substrates.

New insights into the anaerobic degradation of very different categories of compounds, and into process and reactor technology will lead to very promising new generations of anaerobic treatment systems (Lettinga et al., 1997). These concepts will provide a higher efficiency at higher loading rates, are applicable for extreme environmental conditions (e.g. low and high temperatures) and to inhibitory compounds. Moreover, by integrating the anaerobic process with other biological methods (sulfate reduction, micro-aerophilic organisms) and with physical–chemical methods, a complete treatment of the wastewater can be accomplished at very low costs, while at the same time valuable resources can be recovered for reuse.

Anaerobic treatment of chemical and brewery wastewater with a new type of anaerobic reactor, the Biobed® EGSB reactor was reportedly very effective (Zoutberg and Frankin, 1996). The new ultra high loaded type of anaerobic reactor is in full-scale implementation to treat wastewaters from the chemical industry and the brewery. The chemical factory involved is Caldic Europoort in the Netherlands. In this factory formaldehyde is produced from methanol. The wastewater is characterized by high concentrations of these compounds (formaldehyde up to 5 g/L and methanol up to 10 g/L). Due to the special configuration of the employed EGSB reactor, it is possible to acquire removal efficiency for both compounds of more than 99%. At the brewery plant, the Biobed® reactor was installed before an existing aerobic treatment. In this application, the reactor served as a "COD remover", which results in a substantial decreased COD load to the aerobic post-treatment causing lesser sludge production and lesser energy consumption. It is possible to treat wastewater containing toxic but degradable chemical compounds.

The Biobed® EGSB system is able to overcome shortcomings of the upflow anaerobic sludge blanket reactor in the chemical industry (Zoutberg and de Been, 1997). The most striking feature is the growth of biomass in a granular form, similar to the UASB granules: no carrier material is used. The process is especially suitable to treat wastewater that contains compounds that are toxic in high concentrations and that only can be degraded in low concentrations (chemical industry). It is also possible to operate the reactor as an ultra-high loaded anaerobic reactor (to 30 kg COD/m^3·d) for applications in other sectors of the industry (e.g. brewery, yeast, sugar, corn ethanol production etc.).

Psychrophilic (8°C) anaerobic treatment of partly acidified waste water is also applicable using an EGSB system (van Lier et al., 1997) as the average $COD_{soluble}$ and VFA–COD removal efficiencies were 97 and 90%, respectively. Besides, psychrophilic (2 to 20°C) anaerobic treatment of low strength synthetic and malting wastewater was also possible (Rebac et al., 1999). The COD removal efficiencies found in the experiments exceeded 90% in the single module reactor. When a two module EGSB system was used at the temperature range 10–15°C, soluble COD removal and volatile fatty acids removal of 67–78% and 90–96% was achieved, respectively. The mineralization of anthranilic acid as the only carbon and energy source was possible at low influent concentrations (Razo-Flores et al., 1999).

Mesophilic conditions at 35°C to treat slaughterhouse wastewater in an EGSB system appear to be a feasible option (Nunez and Martinez, 1999). The average COD removal efficiency was 67%. Total suspended solid (TSS) removal was 90%. Fats were 85% removed and no accumulation of fats on the sludge was observed. The specific methanogenic activity of the sludge was about 3 times higher than that of the sludge inoculated into the reactor. The sludge activity did not change significantly after one year of operation.

Thermophilic sulfite and sulfate reduction offers good prospects as part of an alternative technology to conventional off-gas desulfurization technologies (Weijma et al., 2000). Methanol can be efficiently used as electron and carbon source to obtain high sulfite and sulfate elimination rates in thermophilic bioreactors.

In Germany, there are currently 125 full-scale anaerobic treatment plants treating industrial wastewater from beet sugar, starch, pectin brewery, distillery, vegetable, and potato processing. The first EGSB reactor at a German potato-processing factory as well as the first municipal wastewater treatment plant combined with a separate anaerobic stage to treat a wastewater mixture from several small factories, demonstrated a successful experience (Austermann-Haun et al., 1999).

A comparison of the behavior of EGSB and UASB reactors in diluted (e.g. ethanol, diluted beer) and concentrated (e.g. coffee) wastewater treatment has been made. There were no big differences in the removal rates during the operation with coffee wastewater. It was likely that in this effluent the process is limited by the reaction kinetics instead of by the mass transfer, due to the complex nature of the waste. With diluted beer, EGSB reactor indicated a better performance than the UASB (Jeison and Chamy, 1999).

The Development of UASB and EGSB

The granular sludge-based UASB and expanded granular sludge bed (EGSB) processes gradually take a large portion of applications. Although UASB remains as the predominant technology in use, EGSB type processes are increasingly gaining more popularity.

Figure 3.1 shows the development regarding changes of process selection in granular sludge systems over time. It is apparent that the traditional UASB system is gradually being replaced by EGSB type systems. This is due to the effectiveness and competitive advantages of the EGSB system.

The data from Fig. 3.2 evidence that the design load for EGSB systems is approximately double that of the UASB process, which results in a

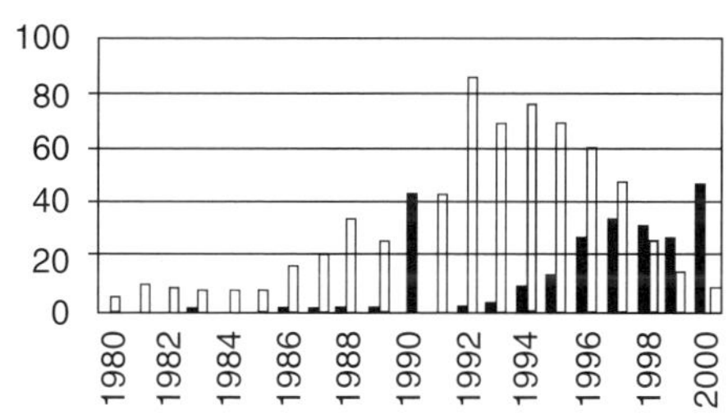

Fig. 3.1. Total number of UASB plants (open bar) and EGSB plants (filled bar) (Frankin, 2001).

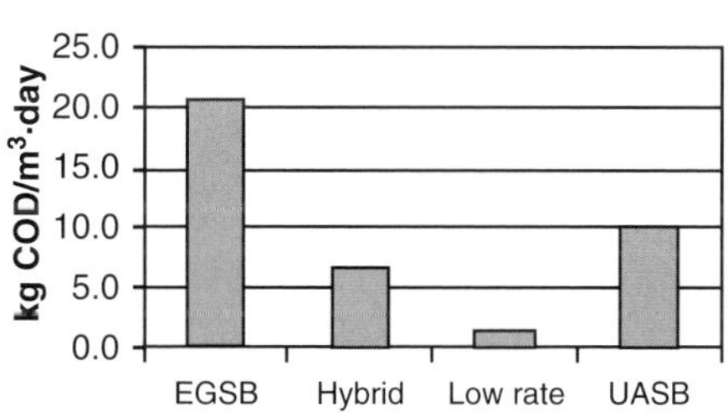

Fig. 3.2. Design loading rates for various processes (Frankin, 2001).

competitive advantage over lower loaded systems. It should however be noted that the data presented represents approximately 50–60% of total anaerobic systems installed and contribution of EGSB and internal circulation (IC) systems may be relatively high in the current database relative to total number of systems installed.

From Fig. 3.2, it follows that the average (design) loading rate for the 198 EGSB plants in the database is somewhat over 20 kg COD/m^3·d. This is two times higher than the average loading rate for 682 UASB designed at 10 kg COD/m^3·d.

The higher design loading rates determine lower cost for reactors, which contributes to the overall cost competitiveness of the process. This explains further the trend as observed in Fig. 3.1.

Hybrid Anaerobic Reactors

Simultaneous removal of trichlorfon, with glucose added as carbon source for degradation requirement of trichlorfon in a hybrid bioreactor was modeled (Chen et al., 1998). The hybrid bioreactor has both suspended biomass and magnetically immobilized biofim. Evaluation of the respective roles of these two types of biomass with a mathematical model was developed and the model also verified well with experimental results. It has been found that the suspended biomass plays a key role in removing both substances in the system. This was due to complete coexistence of both trichlorfon-degrading and glucose-removing bacteria completely in each type of the granules. Such a system is applicable to the treatment of complex industrial wastewaters that contain easily biodegradable organics as well as refractory pollutants.

The use of an upflow anaerobic hybrid blanket (UAHB) reactor, adding water absorbing polymer particles (WAP), to treat a fermentation process wastewater consisting of high sulfate and ammonia, is proven to have a better stability (Imai et al., 1998). The granules were developed in the UAHB process, in which a filter was installed in the upper part of reactor and WAP were also added into inoculum, for treating sulfide- and ammonia-rich wastewater.

Anaerobic thermophilic (55°C) treatment of thermomechanical pulping whitewater (TMP) in reactors based on biomass attachment and entrapment was studied (Jahren et al., 1999) using three different reactor configurations. Up to 70% COD removals were achieved in all the reactors. The anaerobic hybrid reactor, composing a UASB and a filter, exhibited degradation rates up to 10 kg COD/m^3·d. The anaerobic multistage reactor, consisting of three compartments, each packed with granular sludge and carrier elements, showed degradation rates up to 9 kg COD/m^3·d. Clogging and short-circuiting eventually became a problem in the multistage reactor, probably caused by too high packing of the carriers. The anaerobic moving bed biofilm reactor performed similar to the other reactors at loading rates below 1.4 kg COD/m^3·d, which was the highest loading rate applied. The use of carriers in the anaerobic reactors allowed short HRT with good treatment efficiencies for TMP whitewater.

Pentachlorophenol (PCP) is remarkably and efficiently degraded in a hybrid reactor supplied with a mixture of fatty acids (propionic, butyric, acetic, and lactic) and methanol (Montenegro et al., 2001). The reduction of COD was around 97% and methane was found to be 86% in the biogas production. The efficiency of (VFA) breakdown was 93, 64, and 74% respectively for butyric, propionic, and acetic. PCP total removal of more than 99% was reached by granular sludge activities formed during 21 months of reactor operation. Methanogenic microorganisms predominance was noticed with 10^5 to 10^6 cells/mL during enumeration on methanol or lactate added to sulfate culture media. The removal rate was 1.07 mg PCP $\cdot$ g^{-1} Volatile solid (VS) $\cdot$ d^{-1} during the highest PCP concentration addition.

Anaerobic Continuous Stirred Tank Reactor (CSTR)

The most common bioreactor type used for anaerobic digestion is the continuously stirred tank reactor (CSTR). Vanderhaegen et al. (1992) cultivated anaerobic granules in (CSTR), and found that granular sludge disappeared within three weeks when the reactors were incubated statically instead of being shaken. The hydrodynamic shear force appeared to be necessary for maintaining the integrity of granular sludge, and anaerobic granulation may not be reactor type-dependent, but related to the way by which the reactor is operated. It should be pointed out that hydrodynamic flow patterns in UASB and CSTR are quite different. In CSTR, the bacteria move in a stochastic way, i.e. there is no regular liquid upflow pattern. The experiments by Vanderhaegen et al. (1992) challenged the general belief that liquid upflow pattern is essential for the development of granular methanogenic sludge.

Anaerobic Baffled Reactor (ABR)

With the latest development, split feed anaerobic baffled reactor (SFABR) was introduced and it shortened the start-up period and gave higher process performance (Uyanik et al., 2002), when using improved seed material, even for the treatment of particularly problematic wastewater, i.e. ice-cream wastewater. An SEM study revealed that granulation process occured rapidly in the SFABR compared with other reactor configurations, and that the reactor contained a highly mixed population of methanogens in all compartments. The use of polymer-conditioned anaerobic sludge and granular sludge as seed proved advantageous over the use of suspended growth anaerobic sludge, and the "improved" SFABR consequently performed more efficiently and also showed greater stability than the conventional ABR.

Internal Circulation Reactor (IC)

In industrial practice, one major limitation of UASB reactors is the washout of seed sludge during the start-up period. In order to overcome the washout-related operation problem in conventional UASB system, Pereboom and Vereijken (1994) presented a modified design of two-stage UASB reactor, namely the internal circulation (IC) reactor. The IC reactor consists of two inter-connected UASB compartments on top of each other. In the first stage located in the bottom part of the IC reactor, most biogas will be produced, and it will be trapped in the first section of gas-hoods and then rise through the riser section to a gas–liquid separator placed on top of the reactor. The biogas production thus drives an internal circulation flow, which results in excellent mixing in the bottom section. In the second stage in top biomass retention and effluent polishing takes place. It seems that the high turbulence and adequate mixing characteristics make the IC reactor more attractive for reducing clogging and handling high strength organic wastewater. Comparison of results is listed in Table 3.4. Obviously, the performance of IC reactor is comparable with or even better than that of UASB for high strength industrial wastewater treatment.

Table 3.4. The performance of internal circulation (IC) and conventional UASB reactors (extracted from Pereboom and Vereijken, 1994)

	UASB	IC	IC	UASB	IC
Wastewater type	Brewery	Brewery	Brewery	Potato	Potato
Reactor volume (m^3)	1400	50	162 × 6 units	1700 × 2 units	100
VLR (kg COD/m^3·d)	6.8	20	24	10	48
OLR (kg COD/kg VSS·d)	0.2	0.7	0.96	0.35	1.3
Biogas production (m^3/m^3·d)	2.0	5.5		3.0	
COD removal efficiency (%)	80	85	80	95	85
Average diameter of granules (mm)	0.6	0.84	0.66	0.51	0.87

Anaerobic Sequencing Batch Reactor (ASBR)

Anaerobic SBR (ASBR), which has a much simpler configuration than UASB, is a modified design of batch fed process. The major differences between ASBR and UASB are: (a) ASBR does not require feed distribution; (b) there is no three-phase separator in ASBR; and (c) an upflow hydraulic pattern is absent in ASBR system. As compared to UASB system, the non-continuous operation mode of ASBR is the major disadvantage. Wirtz and Dague (1996) successfully cultivated a granular sludge blanket with an ASBR in five months after seeding the reactor with non-granular primary digester sludge. Angenent and Sung (2001) found that MLVSS retained in ASBR is 2.5-fold higher than that in UASB, and the performance in term of COD removal in ASBR is comparable to that of UASB. The research conducted by Kennedy and Lentz (2000) also showed that at low OLRs, the performances of continuous UASB and anaerobic SBR were quite similar, but continuous UASB reactors performed more favorably than the anaerobic SBR at high OLRs.

Anaerobic Migrating Blanket Reactor (AMBR)

Recently, Angenent and Sung (2001) developed a novel anaerobic wastewater treatment system, namely anaerobic migrating blanket reactor (AMBR), which is a continuously fed, compartmentalized reactor without the requirement of elaborate gas–solid-separation and feed-distribution systems. It was found that granules in an AMBR tended to be darker in color, smaller, and denser than granules in a UASB reactor operated under conditions similar to those in the AMBR. The internal structure of AMBR is more complex than that of UASB, especially the multiple-point mechanic mixing to overcome sludge clogging as well as to improve feed distribution. As compared to UASB system, the possible advantages of an AMBR with more than three compartments include smaller biomass migration rates, less chance of short-circuiting, operation in a step feed mode for high-strength wastewater during shock loads, and difficult-to-degrade compounds would be degraded more efficiently.

The Future of Anaerobic Granulation

Anaerobic treatment is well over 100 years old. Its initial development was for the treatment of domestic wastewaters, it then progressed in application to separate sludge digestion, then to treatment of dilute industrial wastewaters. Several processes have been developed that accomplish efficient treatment of wastewaters at short detention times.

The anaerobic granulation system has been known for its unique ability to convert highly objectionable wastes into useful products. With global concerns over energy shortages and greenhouse gas formation through combustion of fossil fuels, more efforts towards renewable energy supplies is clearly needed. Greater efforts are now needed for broader applications of anaerobic granulation system for ridding the environment of unwanted organic materials by converting them into methane, a renewable energy source. The anaerobic granulation process leading towards efficient methane production from wastewaters clearly fits this need. Research towards even broader application is clearly of importance. Problems that need addressing are process reliability, toxicity causes and effects, odor production and control, and better understanding of refractory organic degradation.

From all the numerous and the latest published research on anaerobic processes, cited in the earlier section, it is arguably the most promising wastewater treatment system that is able to meet the desired stringent criteria for future technology in environmentally sustainable development.

Anaerobic granulation process would be the one that is able to minimize environmental harm while increasing industrial productivity and improving quality of life.

At the moment, the most popular treatment process is the UASB reactor. However, with the recent development of EGSB and "Staged Multi-Phase Anaerobic" (SMPA) reactor systems, this may lead to a very promising new generations of anaerobic treatment system (Lettinga et al., 1997). These concepts behind the EGSB will provide a higher efficiency at higher loading rates, are applicable for extreme environmental conditions (e.g. low and high temperatures) and to inhibitory compounds. Moreover, by integrating the anaerobic process with other biological methods (sulfate reduction, micro-aerophilic organisms) and with physical–chemical methods, a complete treatment of the wastewater can be accomplished at very low costs, while at the same time valuable components can be recovered for reuse.

It becomes clear that anaerobic treatment is an established technology for a wide variety of industrial applications. The technology is accepted in the industrialized western world as well as in less developed countries. The granular sludge-based UASB and EGSB processes gradually take a large portion of these applications. Although UASB still is the predominant technology in use, at present EGSB type processes are gaining more popularity driven by economics. The data evidence that the design load for EGSB systems is approximately double that of the UASB process, which results in a competitive advantage over lower loaded systems. It should however be noted that the data presented represents approximately 50–60% of total anaerobic systems installed and contribution of EGSB and IC systems may be relatively high in the current database relative to the total number of systems installed. It is also foreseen that the higher loaded EGSB type systems are gradually replacing at least some of the UASB applications (Frankin, 2001).

In the fields of psychrophilic and thermophilic anaerobic treatment, specific reactor development may contribute to further enhance volumetric conversion capacities. Due to reduced water usage, both COD and salt concentrations tend to increase for industrial effluents. As a consequence, there is a need for the development of anaerobic reactors retaining flocculant or granular biomass. The membrane bioreactors (MBR) offer a solution for certain niches in wastewater treatment (Mulder et al., 2001). However, poor oxygen transfer economy and biomass fouling are major problems of MBR to be overcome. Membrane bioreactors coupled with granular-based anaerobic processes are worth exploring into.

Environmental protection- and resource conservation-concepts focus on pollution prevention and on a minimum of consumptive use of energy, chemicals, and water in pollution abatement and a maximum of re-use of treated wastewater, by-products, and residues produced in the treatment of wastewater. Consequently, by implementing these concepts, wastewaters like sewage and industrial effluents become an important source of water, fertilizers, soil conditioners, and frequently energy instead of a social threat. In addition, a bridge is made between environmental protection and agriculture practice, stimulating urban agriculture in the neighborhood of large cities. Anaerobic granulation process is considered as the core technology for mineralizing organic compounds in highly polluted wastewater streams.

Nowadays, processes based on anaerobic treatment appear to be an excellent option as the core of an integrated process for waste and wastewater treatment (Lema and Omil, 2001). Environmental regulations in the European Union, based on the concept of integrated prevention and control of pollution, are oriented towards the sustainability of the production processes, and this leads to better recovery of resources from raw materials, energy saving, etc. In the last few decades, granular sludge-based anaerobic processes have been receiving widespread acceptance and has been successfully used to treat a variety of industrial wastewaters. The processes offer high degree of organics removal, low sludge production, and low energy consumption along with energy production in the form of biogas. It may not be an unreasonable expectation that, in the future, the treatment technologies will experience a global shift towards usage of highly efficient granular sludge-based anaerobic processes for treatment of wastewaters.

References

Angenent, L.T., & Sung, S. (2001). Development of anaerobic migrating blanket reactor (AMBR), a novel anaerobic treatment system. *Water Research*, *35*, 1739–1747.

Annachhatre, A.P., & Amatya, P.L. (2000). UASB treatment of tapioca starch wastewater. *Journal of Environmental Engineering*, *126* (12), 1149–1152.

Austermann-Haun, U., Meyer, H., Seyfried, C.F., & Rosenwinkel, K.H. (1999). Full-scale experiences with anaerobic/aerobic treatment plants in the food and beverage industry. *Water Science and Technology*, *40* (1), 305–312.

Boardman, G.D., & McVeigh, P.J. (1997). Use of UASB technology to treat crab processing wastewaters. *Journal of Environmental Engineering*, *123* (8), 776–785.

Bogte, J.J., Breure, A.M., van Andel, J.G., & Lettinga, G. (1993). Anaerobic treatment of domestic wastewater in small scale UASB reactors. *Water Science and Technology*, *27* (9), 75–82.

Chen, G.H., Ozaki, H., & Terashima, Y. (1998). Modelling containing simultaneous removal of trichlorfon and glucose in a hybrid bioreactor magnetically immobilized biomass. *Water Science and Technology*, *38* (4–5), 179–186.

Draaijer, H., Maas, J.A.W., Schaapman, J.E., & Khan, A. (1992). Performance of the 5 MLD UASB reactor for sewage treatment at Kanpur, India. *Water Science and Technology*, *25*, 123–133.

Driessen, W.J.B.M., Tielbaard, M.H., & Vereijken, T.L.F.M. (1994). Experience on anaerobic treatment of distillery effluent with the UASB process. *Water Science and Technology*, *30*, 193–201.

Droste, R.L., Kennedy, K.J., Lu, J.G., & Lentz, M. (1998). Removal of chlorinated phenols in upflow anaerobic sludge blanket reactors. *Water Science and Technology*, *38* (8–9), 359–367.

Fang, H.H.P. (1994). Performance and granule characteristics of UASB process treating wastewater with hydrolyzed proteins. *Water Science and Technology*, *30* (8), 55–63.

Field, J.A., Lettinga, G., & Geurts, M. (1987). Methanogenic toxicity and anaerobic degradability of potato starch wastewater phenolic amino acids. *Biological Wastes*, *21* (1), 37–54.

Frankin, R.J. (2001). Full-scale experiences with anaerobic treatment of industrial wastewater. *Water Science and Technology*, *44* (8), 1–6.

Guwy, A.J., Hawjes, F.R., Wilcox, S.J., & Hawkes, D.L. (1997). Neural network and on-off control of bicarbonate alkalinity in a fluidized-bed anaerobic digester. *Water Research*, *31*, 2019–2025.

Harper, S.R., & Suidan, M.T. (1991). Anaerobic treatment kinetics. *Water Science and Technology*, *24*, 61–78.

Hulshoff Pol, L.W., Euler, H., Schroth, S., Wittur, T., & Grohganz, D. (1998). GTZ sectoral project "promotion of anaerobic technology for the treatment of municipal and industrial wastes and wastewater". *In Proceedings of the Fifth Latin-American Seminar on Anaerobic Wastewater Treatment*, Vina del Mar, Chile, October 1998, 27–30.

Hwu, C.S., van Beek, B., van Lier, J.B., & Lettinga, G. (1997a). Thermophilic high-rate anaerobic treatment of wastewater containing long-chain fatty acids: Impact of reactor hydrodynamics. *Biotechnology Letters*, *19*, 447–451.

Hwu, C.S., van Beek, B., van Lier, J.B., & Lettinga, G. (1997b). Thermophilic high-rate anaerobic treatment of wastewater containing long-chain fatty acids: Effect of washed out biomass recirculation. *Biotechnology Letters*, *19*, 453–456.

Hwu, C.S., van Lier, J.B., & Lettinga, G. (1997c). Anaerobic toxicity and degradability of oleic acid under mesophilic and thermophilic conditions. Mededelingen Faculteit Landbouwkundige en Toegepaste Biologische Wetenschappen Universiteit Gent, *62*, 1825–1832.

Imai, T., Ukita, M., Sekine, M., Nakanishi, H., & Fukagawa, M. (1998). Treatment characteristics of high strength fermentation wastewater consisting of high sulfate and ammonia by UAHB process. *Water Science and Technology*, *38* (8–9), 377–384.

Jahren, S.J., Rintala, J.A., & Ødegaard, H. (1999). Anaerobic thermophilic (55°C) treatment of TMP Whitewater in reactors based on biomass attachment and entrapment. *Water Science and Technology*, *40* (11–12), 67–76.

Jeison, D., & Chamy, R. (1999). Comparison of the behaviour of expanded granular sludge bed (EGSB) and upflow anaerobic sludge blanket (UASB) reactors in dilute and concentrated wastewater treatment. *Water Science and Technology*, *40* (8), 91–98.

Kennedy, K.J., & Lentz, E.M. (2000). Treatment of landfill leachate using sequencing batch and continuous upflow anaerobic sludge blanket (UASB) reactors. *Water Research*, *34*, 3640–3656.

Lay, J.J., & Cheng, S.S. (1998). Influence of hydraulic loading rate on UASB reactor treating phenolic wastewater. *Journal of Environmental Engineering*, *124* (9), 829–837.

Lema, J.M., & Omil, F. (2001). Anaerobic treatment: a key technology for a sustainable management of wastes in Europe. *Water Science and Technology*, *44* (8), 133–140.

Lettinga, G. (1995). Anaerobic digestion and wastewater treatment systems. *Antonie van Leeuwenhoek*, *67*, 3–28.

Lettinga, G. (1996). Sustainable integrated biological wastewater treatment. *Water Science and Technology*, *33* (3), 85–98.

Lettinga, G., van der Ben, J., & van der Sar, J. (1976). Anaerobic treatment of sugarbeet waste water (in Dutch). *H_2O*, *9*, 38.

Lettinga, G., Pette, K.C., de Vletter, R., & Wind, E. (1977). Anaerobic treatment of sugarbeet waste water in a 6 m^3 pilot plant (in Dutch). *H_2O*, *10*, 526.

Lettinga, G., van der Geest, A.T., Hobma, S., & Laan, J.V. (1979a). Anaerobic treatment of methanolic wastes. *Water Research*, *13*, 725–738.

Lettinga, G., van Velsen, A.F.M., de Zeeuw, W., & Hobma, S.W. (1979b). Feasibility of the upflow anaerobic sludge blanket process. *In American Society of Civil Engineers*, New York, 35.

Lettinga, G., van Velsen, A.F.M., Hobma, S.W., de Zeeuw, W., & Klapwijk, A. (1980). Use of the upflow sludge blanket reactor concept for biological waste water treatment, especially for anaerobic treatment. *Biotechnology and Bioengineering*, *22*, 699–734.

Lettinga, G., Roersma, R., & Grin, P. (1983). Anaerobic treatment of raw domestic sewage at ambient temperatures using a granular bed upflow anaerobic sludge blanket reactor. *Biotechnology and Bioengineering*, *25*, 1701–1724.

Lettinga, G., Field, J.A., Sierra-Alvarez, R., van Lier, J.B., & Rintala, J. (1991). Future perspectives for the anaerobic treatment of forest industry wastewaters. *Water Science and Technology*, *24* (3–4), 91–102.

Lettinga, G., de Man, A., van der Last, A.R.M., Wiegant, W., van Knippenberg, K., Frijns, J., & van Buuren, J.C.L. (1993). Anaerobic treatment of domestic sewage and wastewater. *Water Science and Technology*, *27* (9), 67–73.

Lettinga, G., van Lier, J., Zeeman, G., & Hulshoff Pol, L.W. (1997). Advanced anaerobic wastewater treatment in the near future. *Water Science and Technology*, *35* (10), 5–12.

Liu, Y., Xu, H.L., Show, K.Y., & Tay, J.H. (2002). Anaerobic granulation technology for wastewater treatment. *World Journal of Microbiology and Biotechnology*, *18* (2), 99–113.

Mamouni, R.El., Leduc, R., & Guiot, S.R. (1998). Influence of synthetic and natural polymers on the anaerobic granulation process. *Water Science and Technology*, *38* (8–9), 341–347.

Marsilli-Libeli, S., & Muller, A. (1996). Adaptive fuzzy pattern recognition in the anaerobic digestion process. *Pattern Recognition Letters*, *17*, 651–659.

Montenegro, M.D.A.P., Moraes, E.D.M., Soares, H.M., & Vazoller, R.F. (2001). Hybrid reactor performance in pentachlorophenol (pcp) removal by anaerobic granules. *Water Science and Technology*, *44* (4), 137–144.

Mulder, R., Vereijken, T.L.F.M., Frijters, C.M.T.J., & Vellingaet, S.H.J. (2001). Future perspectives in bioreactor development. *Water Science and Technology*, *44* (8), 27 32.

Nunez, L.A., & Martinez, B. (1999). Anaerobic treatment of slaughterhouse wastewater in an expanded granular sludge bed (EGSB) reactor. *Water Science and Technology*, *40* (8), 99–106.

Pereboom, J.H.F., & Vereijken, T.L.F.M. (1994). Methanogenic granule development in full scale internal circulation reactors. *Water Science and Technology, 30*, 9–21.

Petruy, R., & Lettinga, G. (1997). Digestion of a milk-fat emulsion. *Bioresource Technology, 61*, 141–149.

Petruy, R., Field, J.A., & Lettinga, G. (1997). Anaerobic biodegradation of a milk-fat emulsion in an expanded granular sludge bed reactor. Mededelingen Faculteit Landbouwkundige en Toegepaste Biologische Wetenschappen Universiteit Gent, *62*, 1833–1840.

Polanco, F.F., Hidalgo, M.D., Polanco, M.F., & Encina, P.A.G. (1999). Anaerobic treatment of polyethylene terephthalate (PET) wastewater from lab to full-scale. *Water Science and Technology, 40* (8), 229–236.

Puñal, A., & Lema, J.M., (1999). Anaerobic treatment of wastewater from a fish-canning factory in a full-scale upflow anaerobic sludge blanket reactor. *Water Science and Technology, 40* (8), 57–62.

Puñal, A., Melloni, P., Roca, E., Rozzi, A., & Lema, J.M. (2001). Automatic start-up of UASB reactors. *Journal of Environmental Engineering, 127* (5), 397–402.

Razo-Flores, E., Smulders, P., Prenafeta-Bold, F., Lettinga, G., & Field, J.A. (1999). Treatment of anthranilic acid in an anaerobic expanded granular sludge bed reactor at low concentrations. *Water Science and Technology, 40* (8), 187–194.

Rebac, S., van Lier, J.B., Lens, P., Stams, A.J.M., Dekkers, F., Swinkels, K.T.M., & Lettinga, G. (1999). Psychrophilic anaerobic treatment of low strength wastewaters. *Water Science and Technology, 39* (5), 203–210.

Rintala, J., Sanz Martin, J.L., & Lettinga, G. (1991). Thermophilic anaerobic treatment of sulfate-rich pulp and paper integrate process water. *Water Science and Technology, 24* (3–4), 149–160.

Rintala, J.A., & Lepisto, S.S. (1997). Pilot-scale thermophilic anaerobic treatment of wastewaters from seasonal vegetable processing industry. *Water Science and Technology, 36* (2–3), 279–285.

Sayed, S., van Campen, L., & Lettinga, G. (1987). Anaerobic treatment of slaughterhouse waste using a granular sludge UASB reactor. *Biological Wastes, 21*, 11–28.

Sayed, S., van der Zanden, J., Wijffels, R., & Lettinga, G. (1988). Anaerobic degradation of the various fractions of slaughterhouse wastewater. *Biological Wastes, 23*, 117–142.

Show, K.Y., Tay, J.H., Yang, L., Wang, Y., & Lua, C.H. (2004a). Effects of stressed loading on start-up and granulation in UASB reactors. *Journal of Environmental Engineering, 130* (7), 743–750.

Show, K.Y., Wang, Y. Foong, S.F., & Tay, J.H. (2004b). Accelerated start-up and enhanced granulation in UASB reactors. *Journal Water Research*, *38* (9), 2293–2304.

Sierra-Alvarez, R., Harbrecht, J., Kortekaas, S., & Lettinga, G. (1990). The continuous anaerobic treatment of pulping wastewaters. *Journal of Fermentation and Bioengineering*, *70*, 119–127.

Syutsubo, K., Harada, H., & Ohashi, A. (1998). Granulation and sludge retainment during start-up of a thermophilic UASB reactor. *Water Science and Technology*, *38* (8–9), 349–357.

Tay, J.H., & Yan, Y.G. (1997). Anaerobic biogranulation as microbial response to substrate adequacy. *Journal of Environmental Engineering*, *123* (10), 1002–1010.

Tay, J.H., & Zhang, X.Y. (1999). Neural fuzzy modeling of anaerobic biological wastewater treatment systems. *Journal of Environmental Engineering*, *125*, 1149–1159.

Tay, J.H., & Zhang, X.Y. (2000). A fast predicting neural fuzzy model for high-rate anaerobic wastewater treatment systems. *Water Research*, *34*, 2849–2860.

Tay, J.H., He, Y.X., & Yan, Y.G. (2001). Improved anaerobic degradation of phenol with supplemental glucose. *Journal of Environmental Engineering*, *127* (1), 38–45.

Ten-Brummeler, E., Pol, L.W.H., Dolfing, J., Lettinga, G., & Zehnder, A.J.B. (1985). Methanogenesis in an upflow anaerobic sludge blanket reactor at pH 6 on an acetate-propionate mixture. *Applied and Environmental Microbiology*, *49*, 1472–1477.

Uyanik, S., Sallis, P.J., & Anderson, G.K. (2002). Improved split feed anaerobic baffled reactor (SFABR) for shorter start-up period and higher process performance. *Water Science and Technology*, *46* (4–5), 223–230.

Vanderhaegen, B., Ysebaert, E., Favere, K., Van Wambeke, M., Peeters, T., Panic, V., Vandenlangenbergh, V., & Verstracte, W. (1992). Acidogenesis in relation to in-reactor granule yield. *Water Science and Technology*, *25*, 21–30.

van der Last, A.R.M., & Lettinga, G. (1992). Anaerobic treatment of domestic sewage under moderate climatic (Dutch) conditions using upflow reactors at increased superficial velocities. *Water Science and Technology*, *25* (7), 167–178.

van Lier, J.B., Rebac, S., & Lettinga, G. (1997). High-rate anaerobic wastewater treatment under psychrophilic and thermophilic conditions. *Water Science and Technology*, *35* (10), 199–206.

van Velsen, A.F.M., Lettinga, G., & den Ottelander, D. (1979). Anaerobic digestion of piggery waste: 3. Influence of temperature. *Netherlands Journal of Agricultural Science*, *27*, 255–267.

Vieira, S.M.M., & Garcia, A.D. (1992). Sewage treatment by UASB-reactor: operation results and recommendations for design and utilization. *Water Science and Technology*, *25*, 143–157.

Vieira, S. M.M., Carvalho, J.L., Barijan, F.P.O., & Rech, C.M. (1994). Application of the UASB technology for sewage treatment in a small community at Sumare, Sao Paulo State. *Water Science and Technology*, *30*, 203–210.

Weijma, J., Haerkens, J.P., Stams, A.J., Hulshoff Pol, L.W., & Lettinga, G. (2000). Thermophilic sulfate and sulfite reduction with methanol in a high rate anaerobic reactor. *Water Science and Technology*, *42* (5–6), 251–258.

Wen, T.C., Cheng, S.S., & Lay J.J. (1995). A kinetic model of a recirculated upflow anaerobic sludge blanket treating phenolic wastewater. *Water Environment Research*, *67* (6), 1005–1006.

Wirtz, R.A., & Dague, R.R. (1996). Enhancement of granulation and start-up in anaerobic sequencing batch reactor. *Water Environment Research*, *68*, 883–892.

Zhou, G.M., & Fang, H.H.P. (1998). Competition between methanogenesis and sulfidogenesis in anaerobic wastewater treatment. *Water Science and Technology*, *38* (8–9), 317–324.

Zoutberg, G.R., & Frankin, R. (1996). Anaerobic treatment of chemical and brewery waste water with a new type of anaerobic reactor: The Biobed® EGSB reactor. *Water Science and Technology*, *34* (5–6), 375–381.

Zoutberg, G.R., & de Been, P. (1997). The Biobed EGSB (Expanded Granular Sludge Blanket) system covers shortcomings of the UASB reactor in the chemical industry. *Water Science and Technology*, *35* (10), 183–188.

Chapter 4

Mechanisms of Aerobic Granulation

Yu Liu

Introduction

As Liu and Tay (2004) noted, many factors including substrate composition, organic loading, hydrodynamic shear force, feast–famine regime, feeding strategy, dissolved oxygen, reactor configuration, solids retention time, cycle time, settling time, and exchange ratio, would influence the formation and properties of aerobic granules developed in sequencing batch reactors (SBR). However, only those parameters associated with selection pressures on the sludge particles would contribute to the formation mechanism of aerobic granules (Liu et al., 2005a). In SBR, two major selection pressures had been identified, i.e. the settling time and the volume exchange ratio (McSwain et al., 2004; Qin et al., 2004a,b; Hu et al., 2005; Liu et al., 2005a; Wang et al., 2005a), while Arrojo et al. (2004) and Wang et al. (2005a) further showed that the discharge time of SBR would be the third important selection pressure of aerobic granulation. In fact, selection pressure has been found to be the key driving force towards successful anaerobic granulation in upflow anaerobic sludge blanket reactor (Hulshoff Pol et al., 1988; Alphenaar et al., 1993). To accelerate industrial application of the aerobic granulation technology, a sound understanding of the mechanisms behind aerobic granulation is highly desirable.

A Generic Four-step Immobilization Mechanism

It has been recognized that aerobic granulation is a process of microbial self-immobilization without the support of carrier (Tay et al., 2001a). Similar to the formation of biofilms and anaerobic granules, aerobic granulation should be a multiple-step process in which a number of physico-chemical and biological factors should be involved. According to the information available for the formation of biofilm and anaerobic granules, Liu and Tay (2002a) proposed a generic four-step formation process for aerobic granulation.

Step 1: Physical movement to initiate bacterium-to-bacterium contact by following forces

- hydrodynamic force;
- diffusion force;
- gravity force;
- thermodynamic forces, e.g. Brownian movement;
- cell mobility. Cells can move by means of flagella, cilia, or pseudopods. Cell mobility had been found to be important for both initial interaction with the surface and movement along the surface in the formation of biofilms (Pratt and Kolter, 1998).

Step 2: Initial attractive forces to keep stable bacterium–bacterium contact, including

- physical forces:
  - van der Waals forces;
  - opposite charge attraction;
  - thermodynamic forces, e.g. free energy of surface, surface tension;
  - hydrophobicity;
  - cross-link or bridge of individual bacteria by filamentous organisms.

In this step, cell surface hydrophobicity would play a crucial role in the initiation of aerobic granules (Tay et al., 2000; Liu et al., 2003, 2004a). According to the thermodynamics theory, increasing the cell surface hydrophobicity would cause a corresponding decrease in the excess Gibbs energy of the surface, which in turn promotes cell-to-cell interaction and further serves as a driving force for bacteria to self-aggregate out of liquid phase (hydrophilic phase). In addition, in this step, filamentous organisms would assist in building up a three-dimensional structure or backbone, which provides a stable environment for the growth of attached bacteria.

- Chemical forces:
    - hydrogen liaison;
    - formation of ionic pairs;
    - formation of ionic triplet;
    - interparticulate bridge and so on.
- Biochemical forces:
    - cellular surface dehydration;
    - cellular membrane fusion.

Tay et al. (2000) postulated that cellular surface dehydration and membrane fusion would play a part in initiating self-immobilization of bacteria, while some environmental conditions would induce cellular surface dehydration and further membrane fusion (Xu et al., 1993).

Step 3: Microbial forces to make aggregated bacteria mature

- production of extracellular polymers, such as exopolysaccharides, etc.;
- growth of cellular cluster;
- metabolic change and genetic competence induced by environment, which facilitate and further strengthen the cell–cell interaction, and finally result in the high density of adhering cells.

Step 4: Stable three-dimensional structure of microbial granules shaped by hydrodynamic shear forces. The microbial granules would be shaped by hydrodynamic shear force to form a certain structured community. The outer shape and size of granules would result from the interactive strength/pattern between granules and hydrodynamic shear force, microbial species and substrate loading rate, and so on. For microbial cells to aggregate, a number of conditions have to be fulfilled. Shear force has been demonstrated to play an important role, and also influences the structure and metabolism of aerobic granules (Tay et al., 2001b; Liu and Tay, 2002). More recently, increasing evidence shows that selection pressure would be the most important factor influencing the formation of aerobic granules (McSwain et al., 2004; Qin et al., 2004a,b; Hu et al., 2005; Liu et al., 2005a; Wang et al., 2005a).

Selection Pressure-driven Aerobic Granulation

Many factors have been known to affect the formation of aerobic granules in sequencing batch reactor (SBR) as briefed earlier. Compared with continuous microbial culture, the main feature of SBR is its cycle operation, i.e. each cycle consists of filling, aeration, settling, and discharging. In SBR, the settling time is likely to exert a selection pressure on the sludge particles. Only particles that can settle down within a given settling time would be retained in the reactor, otherwise they would be washed out of the system. Selection pressure in terms of upflow velocity has been recognized as a driving force towards successful anaerobic granulation in upflow anaerobic sludge blanket (UASB) reactors (Hulshoff Pol et al., 1988; Alphenaar et al., 1993). Similarly, in aerobic granulation a selection pressure should be created to promote the formation of aerobic granules in SBR.

It has been observed that biogranulation can occur by different species including methanogens, acidifying bacteria, nitrifying bacteria, denitrifying bacteria, and aerobic activated sludge. Tay et al. (2002) studied nitrifying granulation and found that even for the defined nitrifying bacteria, nitrifying granules formed only at a strong selection pressure. These seem to indicate that aerobic granulation is a microbial phenomenon induced by environmental conditions through changing the microbial surface properties and metabolic behaviors (Tay et al., 2001b; Pan et al., 2004; Qin et al., 2004b; Wang et al., 2005a). Therefore, aerobic granulation should be species-independent and could be inducible instead of constitutive (Liu et al., 2005a).

Aerobic granulation in SBR could fail without the proper control of settling time or exchange ratio during the operation of SBR. The exchange ratio is defined as the liquid volume withdrawn at the end of the given settling time over the total reactor working volume (Wang et al., 2005a). For a column SBR with the same diameter, the exchange ratio is proportionally correlated to the height from the discharging port to the water surface as illustrated in Fig. 4.1. Settling time and exchange ratio in SBR could be the most effective selection pressures for aerobic granulation.

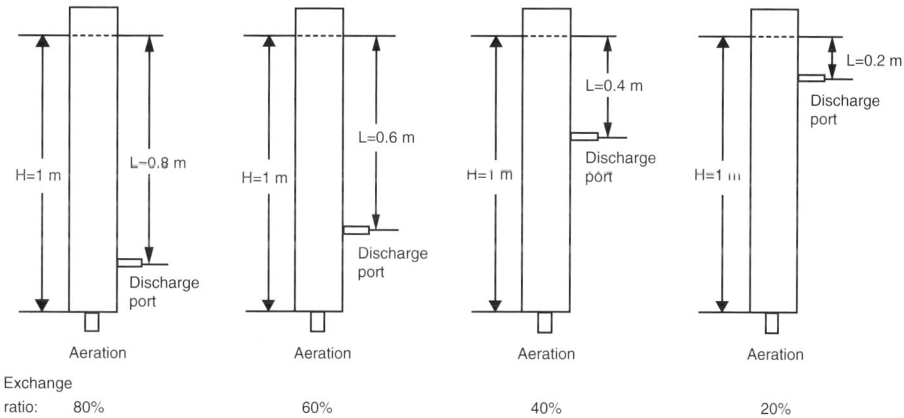

Fig. 4.1. Schematic interpretation of exchange ratios in column SBRs (Liu et al., 2005a).

However, the question is how the settling time and exchange ratio determine aerobic granulation in SBR. In the operation of a column SBR for aerobic granulation, the effluent is discharged at a discharge outlet (Fig. 4.1), i.e. the volume of water above the discharge port is withdrawn at the end of the designed settling time. Liu et al. (2005b) proposed the following equation to describe the settling velocity of bioparticles:

$$V_{\mathrm{s}} = \alpha \frac{d_{\mathrm{p}}^2}{\mathrm{SVI}} e^{-\beta X} \tag{4.1}$$

where V_{s} is the settling velocity of bioparticles, d_{p} is diameter of particle, SVI stands for sludge volume index, X is biomass concentration, α and β are two constant coefficients. Equation (4.1) shows that the settling velocity of aerobic granular sludge is determined by the size of granule, SVI and biomass concentration of granules. If the distance for mixed liquor to travel to the discharge port is L (Fig. 4.1), the corresponding traveling time of bioparticles can be computed as follows

$$\text{Traveling time to the discharge port} = \frac{L}{V_{\mathrm{s}}} \tag{4.2}$$

Equation (4.2) shows that a higher V_{s} results in a shorter traveling time for bioparticles to the discharge port. Hence, the bioparticles with a traveling time longer than the designed settling time would be discharged out of the reactor. In this case, Liu et al. (2005a) proposed that there would be a minimum settling velocity, $(V_s)_{min}$, for bioparticles to be retained in the reactor; and it can be defined as follows:

$$(V_s)_{min} = \frac{L}{\text{settling time}} \tag{4.3}$$

Equation (4.3) implies that bioparticles with a settling velocity less than $(V_s)_{min}$ could be withdrawn from the reactor, while only those bioparticles with a settling velocity greater than $(V_s)_{min}$ can be retained in the system.

As shown in equation (4.3), $(V_s)_{min}$ is a function of the settling time and L, which is proportionally related to the exchange ratio (Fig. 4.1). Therefore, the fastest settling bioparticles are heavy, spherical aggregates, while the slowest settling particles, which sometimes cannot be settled properly, are tiny, light, irregularly shaped aggregates. It appears that bioparticles can be selected according to their settling velocity. Equation (4.3) provides the explanation why the settling time and exchange ratio in SBR can serve as the effective selection pressures that allow the selection of good settling bioparticles, leading towards successful aerobic granulation. It has been proposed that SBR should have a high H/D ratio to improve the selection of granules by the difference in settling velocity (Beun et al., 2002). In fact, it seems that the H/D ratio of SBR is not a selection pressure for aerobic granulation, but a larger H/D ratio is desirable in the design of full-scale SBR because it may allow more space for engineers to manipulate L and subsequently $(V_s)_{min}$ according to actual needs. In addition, there is strong evidence that selection pressures have a profound effect on the surface properties of aerobic granules in terms of cell surface hydrophobicity and extracellular polysaccharides, which in turn favor the formation of aerobic granules in SBR. A similar phenomenon was also observed in anaerobic granulation in UASB reactors (Mahoney et al., 1987; Schmidt and Ahring, 1996). As equation (4.1) shows, the settling velocity of a bioparticle is closely related to the diameter of the aggregate. Thus, it is likely that microbial granulation induced by selection pressures is an effective microbial survival strategy that enables the bacteria to aggregate into big granules and consequently avoid being discharged.

Qin et al. (2004a) studied the effects of various settling times of 5 to 20 min on aerobic granulation at a fixed L, while Wang et al. (2005a) looked into the effects of different exchange ratios on aerobic granulation at a given settling time. Thus, using these data, Liu et al. (2005a) further calculated the minimum settling velocity required for bioparticles to be retained in SBR operated under various settling times and exchange ratios. Figure 4.2 shows the relationship between $(V_s)_{min}$ and the fraction of aerobic granules to total biomass by weight in SBRs. It can be seen that the fraction of aerobic granules to total biomass in the reactor almost linearly increases with the increase in $(V_s)_{min}$, indicated by a correlation coefficient of 0.96.

Figure 4.2 may imply that the effects of settling time and exchange ratio on aerobic granulation can be unified to and interpreted very well by $(V_s)_{min}$, through which good settling bioparticles would be selected and retained in the reactor. When $(V_s)_{min}$ is smaller than 3.8 m h^{-1}, the suspended bioflocs are dominant in the system (Fig. 4.2). In fact, the typical settling velocity of suspended activated sludge is generally less than 4 to 5 m h^{-1} as reviewed by Giokas et al. (2003). Thus, if the SBR is operated at a $(V_s)_{min}$ below 3.8 m h^{-1}, suspended sludge could not be effectively withdrawn from the reactor. Successful aerobic granulation

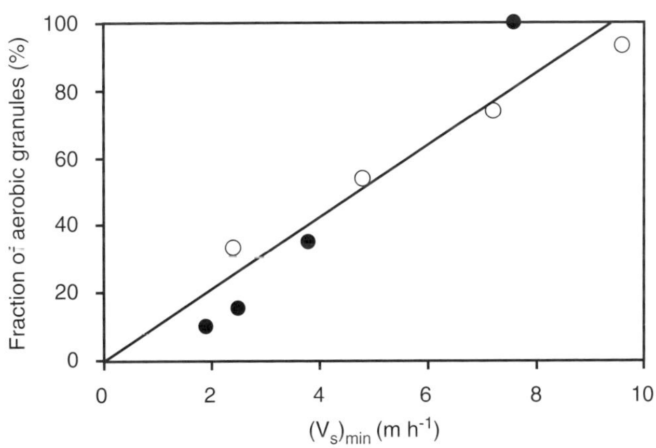

Fig. 4.2. Relationship between the fraction of aerobic granules to total biomass and $(V_s)_{min}$: *filled circles* at different settling times and a constant L of 0.63 m and *hollow circles* at various L and a constant settling time of 5 min (Liu et al., 2005a).

was also reported at the respective settling velocities of 10.0 m h^{-1} and 16.2 m h^{-1} (Beun et al., 2000, 2002). The growth rates of aerobic granules were found to be much lower than that of suspended activated sludge (Yang et al., 2004; Liu et al., 2005c). It should be a reasonable consideration that suspended sludge could easily outcompete aerobic granules due to its faster growth. Such outcompetition in turn would repress aerobic granulation and eventually leads to the disappearance of the aerobic granular sludge blanket in SBR if suspended sludge is not effectively withdrawn. Therefore, Liu et al. (2005a) recommended that $(V_s)_{min}$ must be controlled at a level higher than the settling velocity of suspended sludge, otherwise a rapid and successful aerobic granulation could not be achieved and maintained stably in SBR. Therefore, enhanced selection of bioparticles for rapid aerobic granulation can be realized through properly controlling and adjusting the settling time or the exchange ratio in SBR. However, compared with the exchange ratio, control of the settling time is a more flexible manipulation during full-scale SBR operation (Liu et al., 2005a).

Role of Extracellular Polymeric Substances (EPS) in Aerobic Granulation

Extracellular polymeric substances (EPS) are sticky materials secreted by cells, and may play an important role in cell adhesion phenomena, formation of matrix structure, microbial physiology, and improvement of long-term stability of granules (Schmidt and Ahring, 1994; Tay et al., 2001c; Liu et al., 2004b; McSwain et al., 2005). High polysaccharide content could facilitate cell-to-cell interaction and further strengthen microbial structure through the formation of a polymeric matrix. The accumulation of EPS as capsular material and peripheral slime has been correlated with biological adhesion and aggregation processes (Costerton et al., 1981; Tay et al., 2001c; Liu et al., 2002). The metabolic blocking of exopolysaccharide synthesis was found to prevent microbial aggregation (Cammarota and Sant'Anna, 1998; Yang et al., 2004). EPS in granules were hypothesized to bridge two neighboring bacterial cells physically to each other as well as with other inert particulate matter, and settle out as aggregates (Liu et al., 2004b).

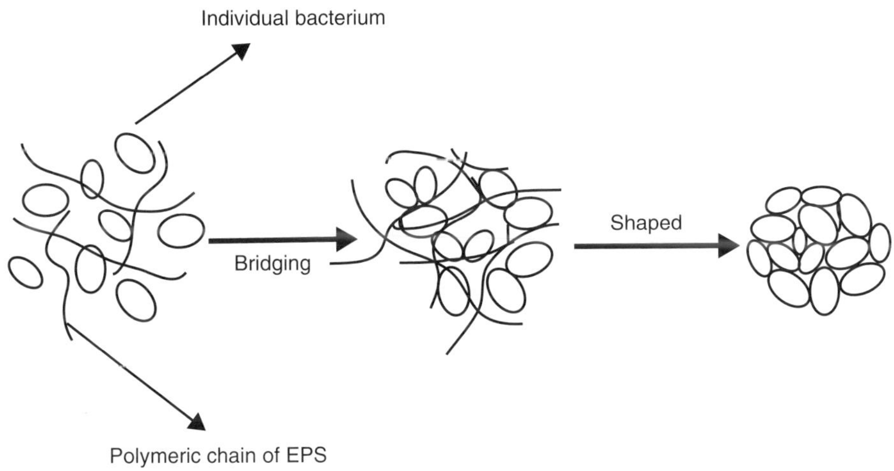

Fig. 4.3. Schematic representation of extracellular polymeric substance-enhanced biogranulation (Liu et al., 2004b).

EPS has been observed in different types of biogranules by scanning electron microscopy and transmission electron microscopy. In the biogranulation process, EPS could provide an extensive surface area for bacterial binding (Fig. 4.3). Furthermore, extracellular polysaccharide matrices surrounding aggregated bacteria can provide sites available for attraction of organic and inorganic materials (Yu et al., 2001; Sponza, 2002; Liu et al., 2004b). Evidence shows that the formation of biogranules is a microbial evolution instead of a random aggregation of suspended microbes (El-Mamouni et al., 1995; Fang, 2000; Tay et al., 2001c).

The spatial distribution of EPS in biogranules should be correlated to microbial evolution and distribution during the formation of biogranules. Investigation into the spatial distribution of EPS with depth in heterotrophic biofilms showed that EPS production yields tended to decrease with biofilm depth (Zhang and Bishop, 2001). This is probably due to the fact that viable biomass loses its ability to produce EPS in the deeper sections of biofilms because of the lower microbial activity resulting from lower nutrient availability. More recently, Wang et al. (2005b) found that the outer shell of aerobic granule was composed of poorly soluble and non-easily biodegradable EPS, whereas its core part was filled with readily soluble and biodegradable EPS. Figure 4.4 shows that the fluorescent

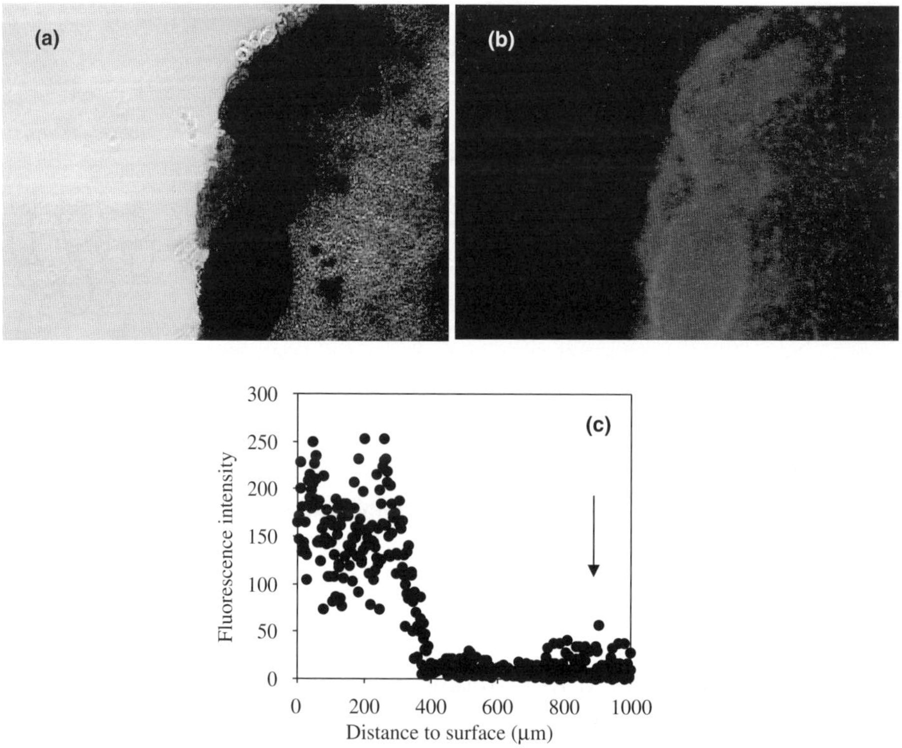

Fig. 4.4. Cross section view of aerobic granules; (a) fresh granule; (b) granule stained by calcofluor white. Bar: 100 μm; (c) profile of the dye fluorescence intensity distribution along the granule radius from the surface to the center (arrow) (Wang et al., 2005b). (See Color Plate Section before the Index.)

dye (calcofluor white) was mainly attached to the outer shell of the granule, while the fluorescence was very weak in the center of the granule. In addition, the EPS produced by bacteria could be utilized as a secondary substrate in the deeper layers or zones of aerobic granules, where readily degradable substrates were either not available or limiting (Chi, 2005; Wang et al., 2005b). It appears that the spatial distribution and properties of EPS rather than its absolute quantity in aerobic granules play an essential role in stabilizing the structure and maintaining the strength of microbial aggregates (Wang et al., 2005b).

It appears from Fig. 4.4 that the fluorescence intensity profile in the direction of granule radius indicates that most calcofluor white-stained

EPS is situated in the outer shell of the granule with a depth of 400 μm below the granule surface. Therefore, β-linked EPS would be mainly located in the outer shell of the granule. It is believed that the insoluble EPS present in the granule shell would play a protective role with respect to the structure stability and integrity of aerobic granules, i.e. insoluble EPS may serve as the backbone of aggregated structure, while the easily biodegradable EPS located at the core of granules would play a less important role.

Summary

It appears that aerobic granulation in SBR is driven by selection pressures exerted on microorganisms. According to the information available so far, the major selection pressures responsible for aerobic granulation in SBR include settling time and volume exchange ratio. It was shown that these major selection pressures indeed could be unified into the concept of minimal settling velocity of bioparticles that determines aerobic granulation in SBR. In addition, EPS seems also to play a certain part in the formation of aerobic granules in SBR.

References

Alphenaar, P.A., Visser, A., & Lettinga, G. (1993). The effect of liquid upflow velocity and hydraulic retention time on granulation in UASB reactors treating wastewater with a high-sulphate content. *Bioresour. Technol.*, *43*, 249–258.

Arrojo, B., Mosquera-Corral, A., Garrido, J.M., & Méndez, R.R. (2004). Aerobic granulation with industrial wastewater in sequencing batch reactors. *Water Res.*, *38*, 3389–3399.

Beun, J.J., van Loosdrecht, M.C.M., & Heijnen, J.J. (2000). Aerobic granulation. *Water Sci. Technol.*, *41*, 41–48.

Beun, J.J., van Loosdrecht, M.C.M., & Heijnen, J.J. (2002). Aerobic granulation in a sequencing batch airlift reactor. *Water Res.*, *36*, 702–712.

Cammarota, M.C., & Sant'Anna, G.L. (1998). Metabolic blocking of exopolysaccharides synthesis: effects on microbial adhesion and biofilm accumulation. *Biotechnol. Lett.*, *20*, 1–4.

Chi, H.Y. (2005). Response of aerobic granules to long-term starvation. *Report of Master of Science*, Nanyang Technological University, Singapore.

Costerton, J.W., Irvin, R.T., & Cheng, K.J. (1981). The bacterial glycocalyx in nature and disease. *Annu. Rev. Microbiol., 35*, 299–324.

El-Mamouni, R., Leduc, R., Costerton, J.W., & Guiot, S.R. (1995). Influence of the microbial content of different precursory nuclei on the anaerobic granulation dynamics. *Water Sci. Technol., 32*, 173–177.

Fang, H.H.P. (2000). Microbial distribution in UASB granules and its resulting effects. *Water Sci. Technol., 42*, 201–208.

Giokas, D.L., Daigger, G.T., Sperling, M., Kim,Y., & Paraskevas, P.A. (2003). Comparison and evaluation of empirical zone settling velocity parameters based on sludge volume index used a unified settling characteristics database. *Water Res., 37*, 3821–3836.

Hu, L., Wang, J., Wen, X., & Qian, Y. (2005). The formation and characteristics of aerobic granules in sequencing batch reactor (SBR) by seeding anaerobic granules. *Process Biochem., 40*, 5–11.

Hulshoff Pol, L.W., Heijnekamp, K., & Lettinga, G. (1988). The selection pressure as a driving force behind the granulation of anaerobic sludge. *Granular Anaerobic Sludge: Microbiology and Technology* (eds. Lettinga, G., Zehnder, A.J.B., Grotenhuis, J.T.C., & Hulshoff Pol, L.W.), Kluwer, Wageningen, 153–161.

Liu, Y., & Tay, J.H. (2002). The essential role of hydrodynamic shear force in the formation of biofilm and granular sludge. *Water Res., 36*, 1653–1665.

Liu, Y., & Tay, J.H. (2004). State of the art of biogranulation technology for wastewater treatment. *Biotechnol. Adv., 22*, 533–563.

Liu, Y., Xu, H.L., Show, K.Y., & Tay, J.H. (2002). Anaerobic granulation technology for wastewater treatment. *World J. Microbiol. Biotechnol., 18*, 99–113.

Liu, Y., Yang, S.F., Liu, Q.S., & Tay, J.H. (2003). The role of cell hydrophobicity in the formation of aerobic granules. *Curr. Microbiol., 46*, 270–274.

Liu, Y., Yang, S.F., Tay, J.H., Liu, Q.S., Qin, L., & Li, Y. (2004a). Cell hydrophobicity is a triggering force of biogranulation. *Enzyme Microb. Technol., 34*, 371–379.

Liu, Y.Q., Liu, Y., & Tay, J.H. (2004b). The effects of extracellular polymeric substances on the formation and stability of biogranules. *Appl. Microbiol. Biotechnol., 65*, 143–148.

Liu, Y., Wang, Z.W., Qin, L., & Tay, J.H. (2005a). Selection pressure-driven aerobic granulation in a sequencing batch reactor. *Appl. Microbiol. Biotechnol., 67*, 26–32.

Liu, Y., Wang, Z.W., Liu, Y.Q., Qin, L., & Tay, J.H. (2005b). A generalized model for settling velocity of aerobic granular sludge. *Biotechnol. Prog.*, *21*, 621–626.

Liu, L.L., Wang, Z.P., Yao, J., Sun, X.J., & Cai, W.M. (2005c). Investigation on the properties and kinetics of glucose-fed aerobic granular sludge. *Enzyme Microb. Technol.*, *36*, 307–313.

Mahoney, E.M., Varangu, L.K., Cairns, W.L., Kosaric, N., & Murray, R.G.E. (1987). The effect of calcium on microbial aggregation during UASB reactor start-up. *Water Sci. Technol.*, *19*, 249–260.

McSwain, B.S., Irvine, R.L., & Wilderer, P.A. (2004). The influence of settling time on the formation of aerobic granules. *Water Sci. Technol.*, *50*, 195–202.

McSwain, B.S., Irvine, R.L., Hausner, M., & Wilderer, P.A. (2005). Composition and distribution of extracellular polymeric substances in aerobic flocs and granular sludge. *Appl. Environ. Microbiol.*, *71*, 1051–1057.

Pan, S., Tay, J.H., He, Y.X., & Tay, S.T.L. (2004). The effect of hydraulic retention time on the stability of aerobically grown microbial granules. *Lett. Appl. Microbiol.*, *38*, 158–163.

Pratt, L.A., & Kolter, R. (1998). Genetic analysis of *E. coli* biofilm formation: roles of flagella, motility, chemotaxis and type I *pili*. *Mol. Microbiol.*, *30*, 285–293.

Qin, L., Tay, J.H., & Liu, Y. (2004a). Selection pressure is a driving force of aerobic granulation in sequencing batch reactors. *Process Biochem.*, *39*, 579–584.

Qin, L., Liu, Y., & Tay, J.H. (2004b). Effect of settling time on aerobic granulation in sequencing batch reactor. *Biochem. Eng. J.*, *21*, 47–52.

Schmidt, J.E., & Ahring, B.K. (1994). Extracellular polymers in granular sludge from different upflow anaerobic sludge blanket (UASB) reactors. *Appl. Microbiol. Biotechnol.*, *42*, 457–462.

Schmidt, J.E., & Ahring, B.K. (1996). Granular sludge formation in upflow anaerobic sludge blanket (UASB) reactors. *Biotechnol. Bioeng.*, *49*, 229–246.

Sponza, D.T. (2002). Extracellular polymer substances and physicochemical properties of flocs in steady- and unsteady-state activated sludge systems. *Process Biochem.*, *37*, 983–998.

Tay, J.H., Xu, H.L., & Teo, K.C. (2000). Molecular mechanism of granulation: I. H+ translocation–dehydration theory. *J. Environ. Eng.*, *126*, 403–410.

Tay, J.H., Liu, Q.S., & Liu, Y. (2001a). Microscopic observation of aerobic granulation in sequential aerobic sludge blanket reactor. *J. Appl. Microbiol.*, *91*, 168–175.

Tay, J.H., Liu, Q.S., & Liu, Y. (2001b). The effects of shear force on the formation, structure and metabolism of aerobic granules. *Appl. Microbiol. Biotechnol.*, *57*, 227–233.

Tay, J.H., Liu, Q.S., & Liu, Y. (2001c). The role of cellular polysaccharides in the formation and stability of aerobic granules. *Lett. Appl. Microbiol.*, *33*, 222–226.

Tay, J.H., Yang, S.F., & Liu, Y. (2002). Hydraulic selection pressure-induced nitrifying granulation in sequencing batch reactors. *Appl. Microbiol. Biotechnol.*, *59*, 332–337.

Wang, Z.W., Liu, Y., & Tay, J.H. (2005a). The role of SBR mixed liquor volume exchange ratio in aerobic granulation. *Chemosphere*, *62*, 767–771.

Wang, Z.W., Liu, Y., & Tay, J.H. (2005b). Distribution of EPS and cell surface hydrophobicity in aerobic granules. *Appl. Microbiol. Biotechnol.*, *69*, 469–473.

Xu, H.L., Jiao, X.M., & Liu, S.S. (1993). Fluorescence measurement of surface dielectric constant of cell membrane. *Acta Biophys. Sin.*, *9*, 234–239.

Yang, S.F., Tay, J.H., & Liu, Y. (2004). Inhibition of free ammonia to the formation of aerobic granules. *Biochem. Eng. J.*, *17*, 41–48.

Yu, H.Q., Tay, J.H., & Fang, H.H.P. (2001). The roles of calcium in sludge granulation during UASB reactor start-up. *Water Res.*, *35*, 1052–1060.

Zhang, X.Q., & Bishop, P.L. (2001). Spatial distribution of extracellular polymeric substances in biofilms. *J. Environ. Eng.*, *127*, 850–856.

Chapter 5

Factors Affecting Aerobic Granulation

Yu Liu

Introduction

Aerobic granulation can be regarded as the gathering together of cells through cell-to-cell immobilization to form a stable, contiguous, multicellular association. These aggregated granules have a compact structure as compared with suspended sludge flocs. Studies showed that aerobic granulation is a gradual process from seed sludge to compact aggregates, further to granular sludge and finally to mature granules as illustrated in Fig. 5.1. Obviously, for cells in a culture to aggregate, a number of factors could affect this process.

Substrate Composition

Aerobic granules can grow on a wide variety of organic substrates in sequencing batch reactor (SBR), including glucose, acetate, ethanol, phenol, particulate organic matter-rich wastewater, soybean-processing wastewater, and both simulated and real municipal wastewater (Morgenroth et al., 1997; Beun et al., 1999; Peng et al., 1999; Jiang et al., 2002; Tay et al., 2002a; Liu et al., 2003; Pan, 2003; Arrojo et al., 2004; Schwarzenbeck et al., 2004; Su and Yu, 2005), while nitrifying and phosphorus-accumulating granules have also been developed

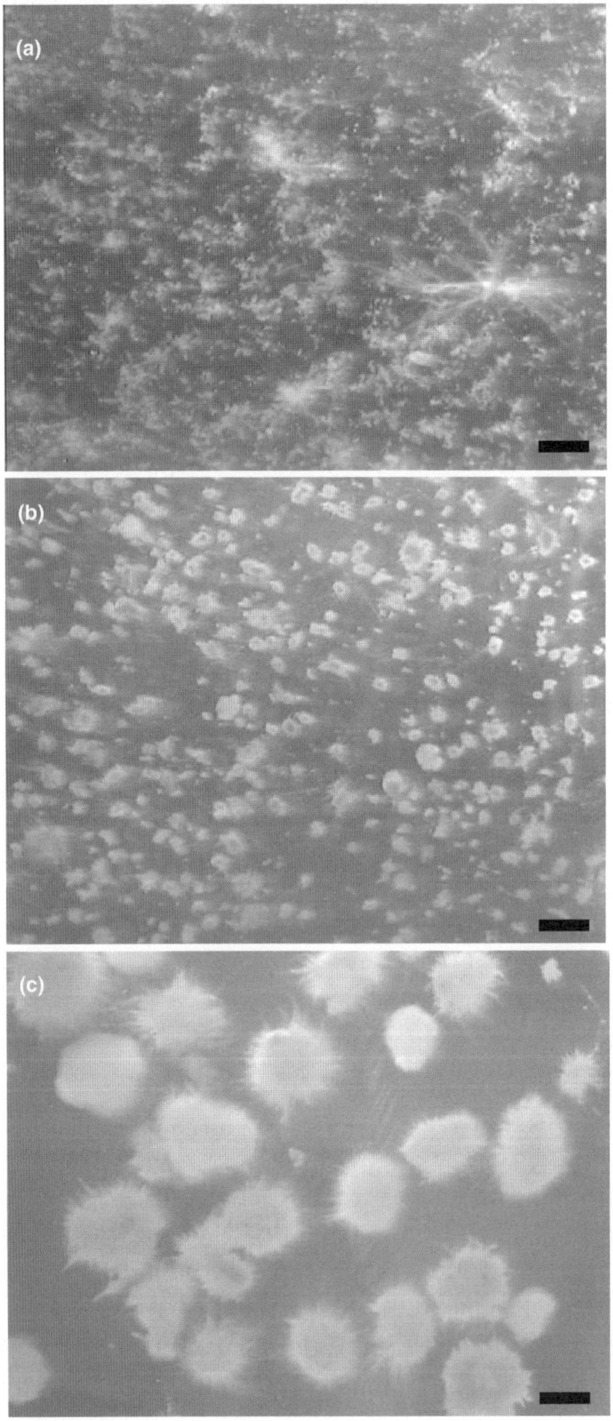

Microbial aggregates formed after 1-week operation of the reactor

Granular sludge formed after 2-week operation of the reactor

Mature granules appeared after 3-week operation of the reactor

Fig. 5.1. Aerobic granulation in the course of operation. Bar: 2 mm (Tay et al., 2001a).

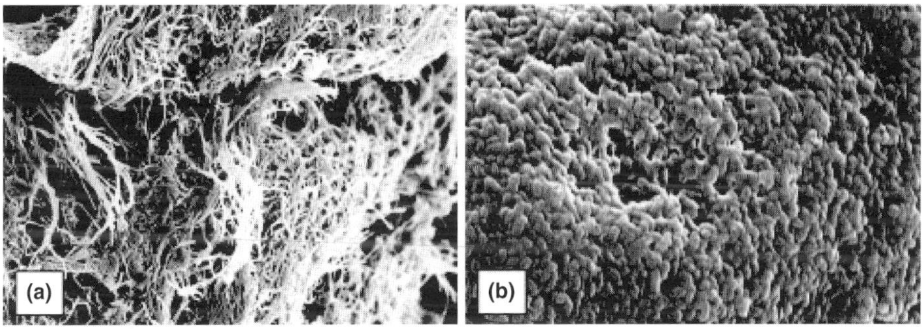

Fig. 5.2. Microstructures of glucose-fed (a) and acetate-fed (b) aerobic granules (Tay et al., 2001a).

(Tay et al., 2002b; Lin et al., 2003; Tsuneda et al., 2004). Figure 5.2 shows that glucose-fed aerobic granules exhibited a filamentous structure, while acetate-fed aerobic granules had a non-filamentous and very compact bacterial structure in which a rod-like species predominated.

These seem to indicate that the formation of aerobic granules is a process independent of or insensitive to the characteristics of the feed wastewater, while evidence shows that the microbial structure and diversity of mature aerobic granules are closely related to the type of substrates used (Tay et al., 2001a, 2002b; Liu et al., 2003).

Substrate Loading Rate

The essential role of organic loading in the formation of anaerobic granules has been recognized, i.e. a relatively high organic loading facilitates the formation of anaerobic granules in the upflow anaerobic sludge blanket (UASB) reactor (Hulshoff et al., 1988; Kosaric et al., 1990). However, this is due to the high organic loading-enhanced biogas production that results in an increased upflow liquid velocity known as the major selection pressure for anaerobic granulation in the UASB reactor (Hulshoff et al., 1988). In contrast to anaerobic granulation, the accumulated evidence shows that aerobic granules can form across an organic loading rate of 2.5–15.0 kg COD m^{-3} day^{-1} (Moy et al., 2002; Liu et al., 2003), while nitrifying granules can also be developed over a very wide range of ammonia–nitrogen loadings (Yang et al., 2003, 2004; Tsuneda et al., 2004;

Qin et al., 2004c). As noted by Liu et al. (2003), aerobic granulation in SBR is substrate concentration-independent, but the kinetics behavior of aerobic granules is related to the applied substrate loading (Moy et al., 2002; Liu et al., 2003). It seems a reasonable consideration that the effect of organic loading rate on the formation of aerobic granules is insignificant, i.e. the substrate loading in the range studied so far is not a determinant of aerobic granulation in SBR. However, the physical characteristics of aerobic granules depend on the organic loading rate (Tay et al., 2003). High loading rate leads to a weakened structure of aerobic granules (Liu et al., 2003; Tay et al., 2003). An increased organic loading rate can raise the biomass growth rate and this in turn reduces the strength of the three-dimensional structure of the microbial community (Liu et al., 2003).

Hydrodynamic Shear Force

In a bubble column SBR, hydrodynamic shear force is mainly created by aeration that can be described by the upflow air velocity. A study showed that higher shear force favored the formation of more compact and denser aerobic granules, while the stimulated production of extracellular polysaccharides and the microbial activity at high shear force was also observed (Fig. 5.3) (Tay et al., 2001b; Liu and Tay, 2002). It is well known that extracellular polysaccharides can mediate both cohesion and adhesion of cells and play a crucial role in maintaining structural integrity

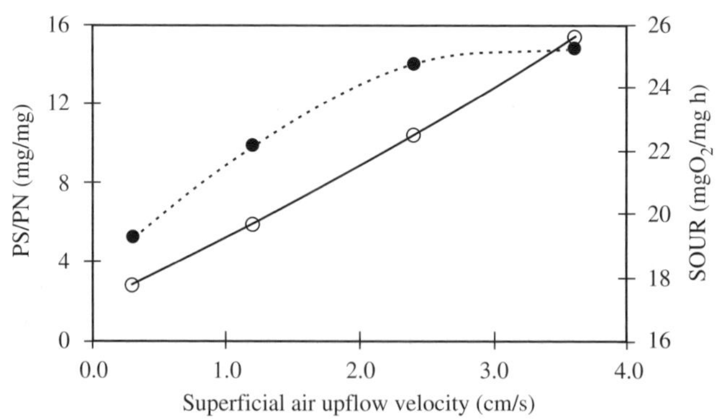

Fig. 5.3. Effects of superficial air upflow velocity on the PS/PN ratio and SOUR of aerobic granules (Tay et al., 2001b). (•) PS/PN; (○) SOUR.

in a community of immobilized cells (Liu et al., 2004b). Consequently, the enhanced production of extracellular polysaccharides at high shear force can make granule structure more compact and stronger.

Similar to the formation of a biofilm, aerobic granules can form at different levels of hydrodynamic shear forces. Therefore, hydrodynamic shear force is not a primary inducer of aerobic granulation in SBR (Liu and Tay, 2002). However, the structure of mature aerobic granules is hydrodynamic shear force-related. High shear in terms of superficial upflow air velocity could lead to more compact, denser, rounder, stronger, and smaller aerobic granules, as showed in Figs 5.4 and 5.5 (Tay et al., 2004).

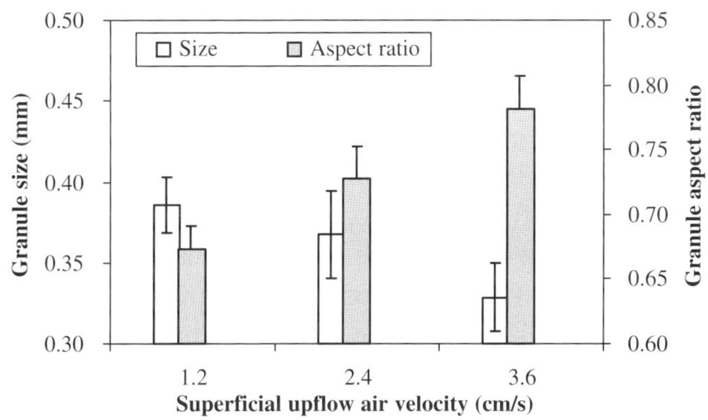

Fig. 5.4. Effect of superficial upflow air velocity on granule size and aspect ratio (Tay et al., 2004).

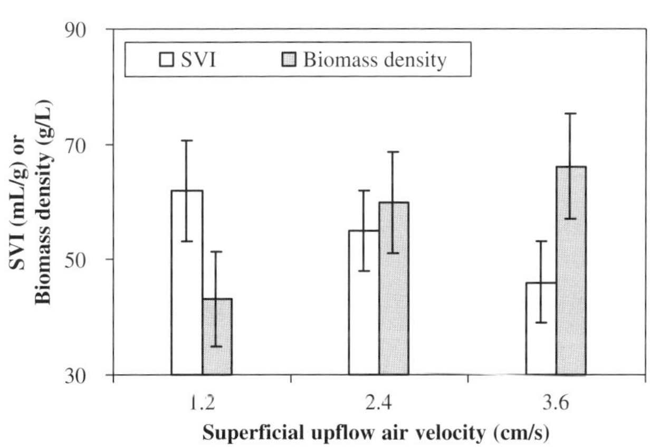

Fig. 5.5. Effect of superficial upflow air velocity on SVI and biomass density (Tay et al., 2004).

Feast–Famine Regime

SBR for cultivation of aerobic granules is operated in a sequencing cycle of feeding, aeration, settling, and discharging of supernatant. In SBR, the aeration period actually consists of two phases: a degradation phase in which the substrate is depleted to a minimum, followed by an aerobic starvation phase in which the external substrate is no longer available. Thus, it is likely that microorganisms in SBR are subjected to a periodic feast and famine regime, called periodic starvation (Tay et al., 2001a). Under the periodic feast–famine conditions, bacteria becomes more hydrophobic and high cell hydrophobicity in turn facilitates microbial aggregation (Bossier and Verstraete, 1996; Tay et al., 2001a; Liu et al., 2004a).

When bacteria are subjected to a periodic feast–famine regime, microbial aggregation could be an effective strategy for cells against starvation. In fact, the periodic feast–famine regime in SBR can be regarded as a kind of microbial selection pressure that may alter the surface properties of cells. However, research also showed that aerobic granules could not be successfully developed if the settling time in SBR was not properly controlled, even though a periodic feast–famine regime was present (Qin et al., 2004a,b), while negative effects of nutrient starvation on the surface properties of aerobic granules in terms of cell hydrophobicity and the content of extracellular polysaccharides were also observed (Zhou, 2004). In addition, when the starvation time in SBR was reduced from 3 h to below 30 min, no significant impact on aerobic granules was observed. This may imply that the periodic feast–famine regime could favor aerobic granulation, but so far there is no solid experimental evidence to show that starvation acts as an inducing force of aerobic granulation in SBR.

Solids Retention Time

Solids retention time (SRT), so-called sludge age, is one of the design parameters in a continuously activated sludge process. Pan (2003) investigated the role of SRT in aerobic granulation in SBR, and found that the sludge age of 20 days or longer is favorable for the formation and maintenance of stable aerobic granules with good settleability and activity (Fig. 5.6). As noted by Liu and Tay (2002), in more than 100 years of

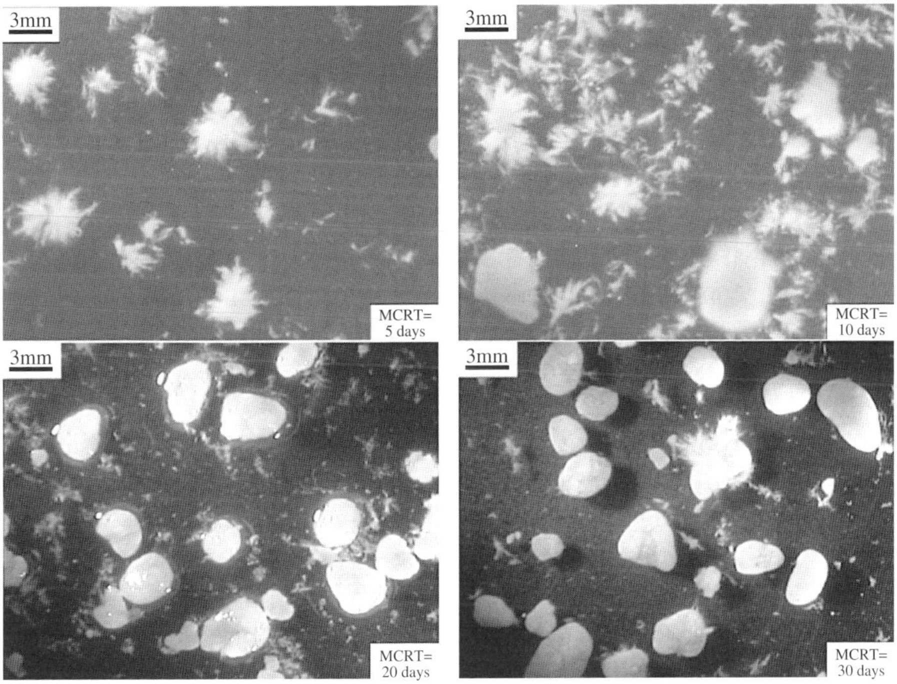

Fig. 5.6. Microscope images of granules/flocs at different SRTs (Pan, 2003).

research history of the activated sludge process, aerobic granulation has never been reported in activated sludge processes operated in an extremely wide range of SRT, from several hours to hundreds of days. It seems reasonable to think that SRT would not be a triggering or inducing force for aerobic granulation.

Dissolved Oxygen

Dissolved oxygen (DO) concentration is an important parameter in the operation of aerobic wastewater treatment systems. Aerobic granules can form at a DO concentration as low as 0.7–1.0 mg l^{-1} in an SBR (Peng et al., 1999), while studies show they can also successfully develop at DO concentrations of 2–6 mg l^{-1} (Yang et al., 2003; Tsuneda et al., 2004; Qin et al., 2004a). It appears that if aerobic condition is maintained by sufficient aeration, the DO concentration is not a decisive parameter in aerobic granulation.

Feeding Strategy

McSwain et al. (2004b) developed an operating strategy to enhance aerobic granulation by intermittent feeding, i.e. different filling times were applied to SBR reactors, resulting in different degrees of feast–famine period for the microorganisms. It was found that a high feast–famine ratio or pulse feeding to the SBR was favorable for the formation of compact and dense aerobic granules, i.e. the feeding strategy may influence the characteristics of aerobic granules formed in a SBR, but it is unlikely to play a role as the trigger for aerobic granulation.

Cycle Time

Basically, the SBR cycle time represents the duration of the substrate oxidation reaction; and it is interrelated to the hydraulic retention time. If the SBR is run at an extremely short cycle time, microbial growth can be hindered by an insufficient reaction time for microorganisms to break down substrates.

As a result, the sludge loss due to hydraulic washout from the system cannot be compensated by the growth of bacteria. Research showed that when a complete washout of sludge blanket occurred, a failure of nitrifying granulation resulted at a very short cycle time (Tay et al., 2002b). As showed in Fig. 5.7, only at a cycle time of 3 h, good nitrifying granules could be formed, while cycle time of 6 and 12 h has no granule formation.

Consequently, if the cycle time is kept much longer than that required for a biological reaction, hydrolysis of the biomass would eventually cause a negative effect on microbial aggregation (Tay et al., 2002b; Chen et al., 2003; Pan et al., 2004; Zhou, 2004). The cycle time of SBR should be short to suppress biomass hydrolysis, but long enough for biomass growth and accumulation in the system. Other research also demonstrated that, for SBRs operated at the optimum cycle time, aerobic granules still could not form if the settling time was kept longer than 15 min (Qin et al., 2004a,b).

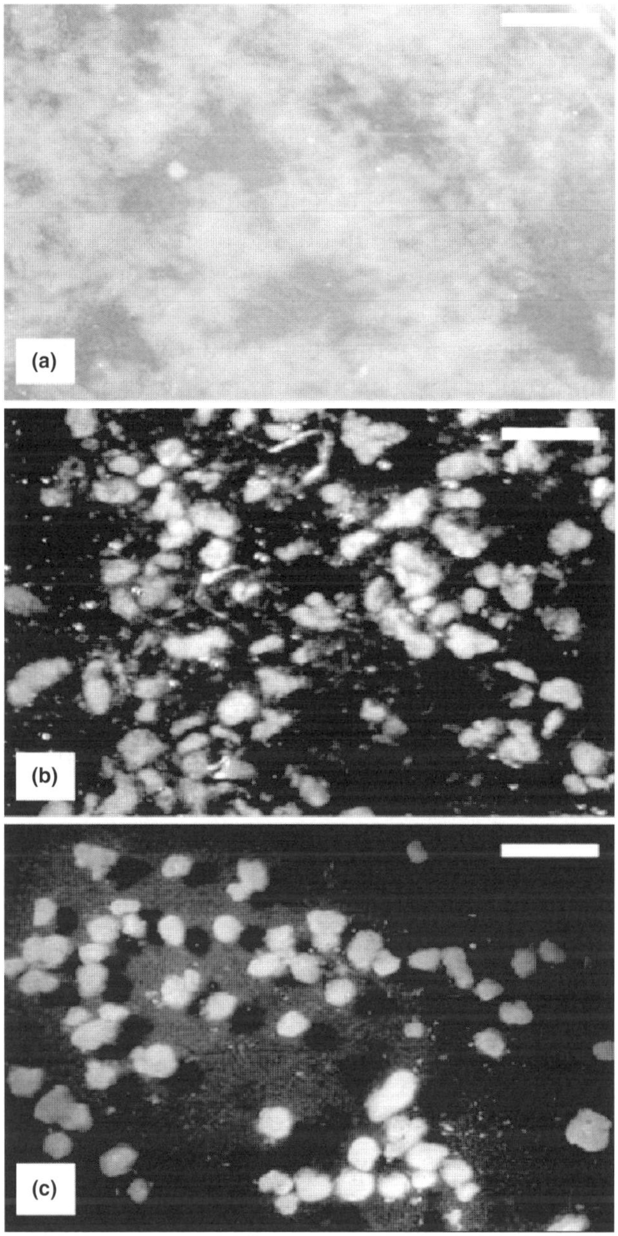

Fig. 5.7. Nitrifying granulation at different cycle time. (a) cycle time of 12 h; (b) 6 h; (c) 3 h. Bar: 1 mm (Tay et al., 2002b).

This implies that cycle time would not be decisive in aerobic granulation in SBR.

Settling Time

In a column SBR, wastewater is treated in successive cycles of a few hours each. At the end of a cycle, settling of the biomass takes place before the effluent is withdrawn. Sludge that cannot settle down within the given settling time could be washed out of the reactor through a fixed discharge port. Basically, a short settling time preferentially selects for the growth of good settling bioparticles. Thus, the settling time exerts a major selection pressure on the microbial community.

Qin et al. (2004a,b) studied the effect of settling time on aerobic granulation in SBR designed with a fixed discharge port, i.e. fixed exchange ratio, and found that aerobic granules were successfully cultivated and became dominant only in SBR operating at a settling time of less than 5 min, while a mixture of aerobic granules and suspended sludge developed in SBR run at longer settling times. In aerobic granulation, a short settling time has been commonly employed to enhance aerobic granulation in SBR (Jiang et al., 2002; Lin et al., 2003; Liu et al., 2003, 2005; Yang et al., 2003; McSwain et al., 2004a; Hu et al., 2005). In fact, at a long settling time, poorly settling sludge flocs cannot be withdrawn effectively; and they may in turn outcompete granule-forming bioparticles. As a result, aerobic granulation could fail in SBR run at longer settling times. This seems to indicate that aerobic granules can form only at short settling times below a critical level, i.e. settling time is a decisive factor in the formation of aerobic granules in SBR. Thus, choice of an optimal settling time is very important in aerobic granulation.

Exchange Ratio

The exchange ratio in SBR is defined as the liquid volume withdrawn at the end of the given settling time over the total reactor working volume. Wang et al. (2005) studied aerobic granulation in SBR run at different exchange ratios in the range of 20–80%, while the settling time was kept constant at 5 min. Thus, a larger exchange ratio is associated with a higher reactor height for discharge. The fraction of aerobic granules in the total biomass was found to be proportionally related to the exchange ratio, e.g. only in the SBR run at the higher exchange ratios of 60 and 80% were aerobic granules dominant; and a mixture of aerobic granules and suspended sludge instead of pure aerobic granules developed at the smaller exchange ratios of 40 and 20%. In addition, Zhu and Wilderer (2003) successfully developed aerobic granules at a fixed exchange ratio of 75%. These results provide experimental evidence that aerobic granulation is highly dependent on the exchange ratio of SBR.

Presence of Calcium Ion in Feed

Jiang et al. (2003) reported that addition of Ca^{2+} accelerated the aerobic granulation process. With addition of 100 mg Ca^{2+} l^{-1}, the formation of aerobic granules took 16 days compared to 32 days in the culture without Ca^{2+} added. The Ca^{2+}-augmented aerobic granules also showed better settling and strength characteristics and had higher polysaccharides contents. It has been proposed that Ca^{2+} binds to negatively charged groups present on bacterial surfaces and extracellular polysaccharides molecules and thus acts as a bridge to promote bacterial aggregation. Polysaccharides play an important role in maintaining the structural integrity of biofilms and microbial aggregates, such as aerobic granules, as they are known to form a strong and sticky non-deformable polymeric gel-like matrix.

Seed Sludge

Aerobic granular sludge SBRs have been seeded with conventional activated sludge. In anaerobic granulation, there is evidence that the characteristics of the seed sludge profoundly influence the formation and properties of anaerobic granules. The important factors that determine the quality of seed sludge for aerobic granulation appear to include the macroscopic characteristics, settleability, surface properties (a high surface hydrophobicity and low surface charge density are preferred), and microbial activity. Although the role of seed sludge in aerobic granulation is not clear yet, some recent research had looked into this aspect (Tay et al., 2005a,b).

Reactor Configuration

Aerobic granules have been produced only in column SBR so far. It can be understood that reactor configuration has an impact on the flow patterns of liquid and microbial aggregates in the reactor. In a column SBR, air flow is subject to an upflow pattern. The air or liquid upflow pattern in a column reactor creates a relatively homogenous circular flow and localized vortex along the reactor height; and thus microbial aggregates are constantly subjected to circular hydraulic attrition (Liu and Tay, 2002).

The feasibility and efficiency of other types of bioreactors, such as completely mixed tank reactor (CMTR) in development of aerobic granular sludge have not been sufficiently demonstrated so far. In a hydrodynamic sense, column-type upflow reactor and CMTR have very different hydrodynamic behaviors in terms of interactive patterns between flow and microbial aggregates, as illustrated in Fig. 5.8a. According to the thermodynamics, the circular flow could force microbial aggregates to be shaped as regular granules that have a minimum surface free energy, provided those aggregates could be kept in the reactors under given dynamic conditions. Thermodynamically, such a phenomenon is very similar to the formation of benthic round-shape boulders in a natural flowing river system. It is obvious that in a column-type upflow reactor a higher ratio of reactor height to diameter can ensure a longer circular flowing trajectory, which in turn creates a more effective hydraulic attrition to microbial aggregates. However, in CMTR microbial aggregates stochastically move with dispersed flow in all directions. Thus, microbial aggregates are

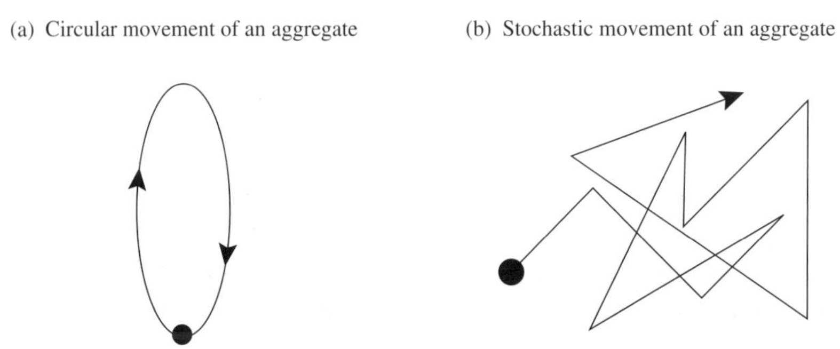

Fig. 5.8. Flow patterns in upflow column reactor (a) and completely mixed tank reactor (b) (Liu and Tay, 2002).

subject to varying localized hydrodynamic shear force, flowing trajectory and random collision (Fig. 5.8b). Under such circumstances, only flocs of irregular shape and size instead of regular granules occasionally form, and this is exactly like what happens in a conventional activated sludge aeration tank, which is a typical CMTR. The operation practice of conventional activated sludge process supports the above analysis because microbial granulation has hardly been reported in CMTR in the past one hundred years of operation practice.

It seems certain that not only the strength of hydrodynamic shear force, but also the interactive pattern between flow and microbial aggregates have effects on the formation of granular sludge. In this aspect, the column-type upflow reactor with high ratio of reactor height to diameter can provide an optimal interactive pattern between flow and microbial aggregates for granulation. It can ensure a circular flowing trajectory, which in turn creates a more effective hydraulic attrition for microbial aggregates. A high *H/D* ratio may also improve oxygen transfer and could result in a reactor with a small footprint (Beun et al., 2002). This may be a major reason why almost all of the granular sludge only forms in column-type upflow reactors. In an engineering sense, the desirable interactive pattern between flow and aggregates might be achieved by controlling reactor configurations and operation strategy. Consequently, a better understanding of the role of flow pattern in granulation process would lead to the development of novel types of granular sludge reactor. On the other hand, some researches also showed that aerobic granules could be developed in SBRs with different *H/D* ratios (Pan, 2003). It seems that aerobic granulation might not tightly associate with the *H/D* ratio.

Summary

This chapter briefly reviewed the factors that may influence or involve in aerobic granulation in SBR, such as substrate composition, substrate loading rate, hydrodynamic shear force, feast–famine regime, solids retention time, dissolved oxygen, feeding strategy, cycle time or hydraulic retention time, settling time, exchange ratio, presence of calcium ion in feed, seed sludge, and reactor configuration. It is expected that such information would be useful in designing aerobic granular sludge SBR for wastewater treatment.

References

Arrojo, B., Mosquera-Corral, A., Garrido, J.M., & Méndez, R.R. (2004). Aerobic granulation with industrial wastewater in sequencing batch reactors. *Water Res.*, *38*, 3389–3399.

Beun, J.J., Hendriks, A., van Loosdrecht, M.C.M., Morgenroth, E., Wilderer, P.A., & Heijnen, J.J. (1999). Aerobic granulation in a sequencing batch reactor. *Water Res.*, *33*, 2283–2290.

Bossier, P., & Verstraete, W. (1996). Triggers for microbial aggregation in activated sludge? *Appl. Microbiol. Biotechnol.*, *45*, 1–6.

Chen, G.H., An, K.J., Saby, S., Brois, E., & Djafer, M. (2003). Possible cause of excess sludge reduction in an oxic-settling-anaerobic activated sludge process (OSA process). *Water Res.*, *37*, 3855–3866.

Hu, L., Wang, J., Wen, X., & Qian, Y. (2005). The formation and characteristics of aerobic granules in sequencing batch reactor (SBR) by seeding anaerobic granules. *Process Biochem.*, *40*, 5–11.

Hulshoff Pol, L.W., Heijnekamp, K., & Lettinga, G. (1988). The selection pressure as a driving force behind the granulation of anaerobic sludge. *Granular Anaerobic Sludge: Microbiology and Technology* (eds. Lettinga, G., Zehnder, A.J.B., Grotenhuis, J.T.C., & Hulshoff Pol, L.W.), Kluwer, Wageningen, 153–161.

Jiang, H.L., Tay, J.H., & Tay, S.T.L. (2002). Aggregation of immobilized activated sludge cells into aerobically grown microbial granules for the aerobic biodegradation of phenol. *Lett. Appl. Microbiol.*, *35*, 439–445.

Kosaric, N., Blaszczyk, R., Orphan, L., & Valladares, J. (1990). The characteristics of granules from upflow anaerobic sludge blanket reactors. *Water Res.*, *24*, 1473–1477.

Lin, Y.M., Liu, Y., & Tay, J.H. (2003). Development and characteristics of phosphorus-accumulating granules in sequencing batch reactor. *Appl. Microbiol. Biotechnol.*, *62*, 430–435.

Liu, Y., & Tay, J.H. (2002). The essential role of hydrodynamic shear force in the formation of biofilm and granular sludge. *Water Res.*, *36*, 1653–1665.

Liu, Q.S., Tay, J.H., & Liu, Y. (2003). Substrate concentration-independent aerobic granulation in sequential aerobic sludge blanket reactor. *Environ. Technol.*, *24*, 1235–1243.

Liu, Y., Yang, S.F., Tay, J.H., Liu, Q.S., Qin, L., & Li, Y. (2004a). Cell hydrophobicity is a triggering force of biogranulation. *Enzyme Microb. Technol.*, *34*, 371–379.

Liu, Y.Q., Liu, Y., & Tay, J.H. (2004b). The effects of extracellular polymeric substances on the formation and stability of biogranules. *Appl. Microbiol. Biotechnol.*, *65*, 143–148.

Liu, Y., Wang, Z.W., & Qin, L. (2005). Selection pressure-driven aerobic granulation in a sequencing batch reactor. *Appl. Microbiol. Biotechnol.*, *67*, 26–32.

McSwain, B.S., Irvine, R.L., & Wilderer, P.A. (2004a). The influence of settling time on the formation of aerobic granules. *Water Sci. Technol.*, *50*, 195–202.

McSwain, B.S., Irvine, R.L., & Wilderer, P.A. (2004b). Effect of intermittent feeding on aerobic granule structure. *Water Sci. Technol.*, *49*, 19–25.

Morgenroth, E., Sherden, T., van Loosdrecht, M.C.M., Heijnen, J.J., & Wilderer, P.A. (1997). Aerobic granular sludge in a sequencing batch reactor. *Water Res.*, *31*, 3191–3194.

Moy, B.Y.P., Tay, J.H., Toh, S.K., Liu, Y., & Tay, S.T.L. (2002). High organic loading influences the physical characteristics of aerobic sludge granules. *Lett. Appl. Microbiol.*, *34*, 407–412.

Pan, S. (2003). Inoculation of microbial granular sludge under aerobic conditions. *Ph.D. Thesis.* Nanyang Technological University, Singapore.

Pan, S., Tay, J.H., He, Y.X., & Tay, S.T.L. (2004). The effect of hydraulic retention time on the stability of aerobically grown microbial granules. *Lett. Appl. Microbiol.*, *38*, 158–163.

Peng, D., Bernet, N., Delgenes, J.P., & Moletta, R. (1999). Aerobic granular sludge—a case report. *Water Res.*, *33*, 890–893.

Qin, L., Tay, J.H., & Liu, Y. (2004a). Selection pressure is a driving force of aerobic granulation in sequencing batch reactors. *Process Biochem.*, *39*, 579–584.

Qin, L., Liu, Y., & Tay, J.H. (2004b). Effect of settling time on aerobic granulation in sequencing batch reactor. *Biochem. Eng. J.*, *21*, 47–52.

Qin, L., Yang, S.F., Tay, J.H., & Liu, Y. (2004c). Aerobic granulation under alternating aerobic and anaerobic conditions in sequencing batch reactors. *Water Environmental Management Series* (eds. Lens, P., & Stuetz, R.), IWA, London, 3–10.

Schwarzenbeck, N., Erley, R., & Wilderer, P.A. (2004). Aerobic granular sludge in a SBR-system treating wastewater rich in particulate matter. *Water Sci. Technol.*, *49*, 41–46.

Su, K.Z., & Yu, H.Q. (2005). Formation and characterization of aerobic granules in a sequencing batch reactor treating soybean-processing wastewater. *Environ. Sci. Technol.*, *39*, 2818–2827.

Tay, J.H., Liu, Q.S., & Liu, Y. (2001a). Microscopic observation of aerobic granulation in sequential aerobic sludge blanket reactor. *J. Appl. Microbiol.*, *91*, 168–175.

Tay, J.H., Liu, Q.S., & Liu, Y. (2001b). The effects of shear force on the formation, structure and metabolism of aerobic granules. *Appl. Microbiol. Biotechnol.*, *57*, 227–233.

Tay, J.H., Liu, Q.S., & Liu, Y. (2002a). Characteristics of aerobic granules grown on glucose and acetate in sequential aerobic sludge blanket reactors. *Environ. Technol.*, *23*, 931–936.

Tay, J.H., Yang, S.F., & Liu, Y. (2002b). Hydraulic selection pressure-induced nitrifying granulation in sequencing batch reactors. *Appl. Microbiol. Biotechnol.*, *59*, 332–337.

Tay, J.H., Pan, S., He, Y.X., & Tay, S.T.L. (2003). Effect of organic loading rate on aerobic granulation: Part II. Characteristics of aerobic granules. *J. Environ. Eng.*, *130*, 1102–1109.

Tay, J.H., Liu, Q.S., & Liu, Y. (2004). The effect of upflow air velocity on the structure of aerobic granules cultivated in a sequencing batch reactor. *Water Sci. Technol.*, *11/12*, 35–40.

Tay, S.T.L., Moy, B.Y.P., Jiang, H.L., & Tay, J.H. (2005a). Rapid cultivation of stable aerobic phenol-degrading granules using acetate-fed granules as microbial seed. *J. Biotechnol.*, *115*, 387–395.

Tay, S.T.L., Moy, B.Y.P., Maszenan, A.M., & Tay, J.H. (2005b). Comparing activated sludge and aerobic granules as microbial inocula for phenol biodegradation. *Appl. Microbiol. Biotechnol.*, *67*, 708–713.

Tsuneda, S., Nagano, T., Hoshino, T., Ejiri, Y., Noda, N., & Hirata, A. (2004). Characterization of nitrifying granules produced in an aerobic upflow fluidized bed reactor. *Water Res.*, *37*, 4965–4973.

Wang, Z.W., Liu, Y., & Tay, J.H. (2005). The role of SBR mixed liquor volume exchange ratio in aerobic granulation. *Chemosphere*, *62*, 767–771.

Yang, S.F., Liu, Y., & Tay, J.H. (2003). A novel granular sludge sequencing batch reactor for removal of organic and nitrogen from wastewater. *J. Biotechnol.*, *106*, 77–86.

Yang, S.F., Liu, Q.S., Tay, J.H., & Liu, Y. (2004). Growth kinetics of aerobic granules developed in sequencing batch reactors. *Lett. Appl. Microbiol.*, *38*, 106–112.

Zhou, J.Q. (2004). Contribution of cell starvation to aerobic granulation. *Report of Master of Science*, Nanyang Technological University, Singapore.

Zhu, J., & Wilderer, P.A. (2003). Effect of extended idle conditions on structure and activity of granular activated sludge. *Water Res.*, *37*, 2013–2018.

Chapter 6

Structure of Aerobically Grown Microbial Granules

Volodymyr Ivanov

Natural Microbial Granules

There are several examples of microbial granules produced in nature. The best-known example is kefir grains, which are the aggregates of lactobacilli, acetobacteria, and yeasts, used in milk fermentation from ancient times. Diameter of these granules range from 1 to 5 mm. It is considered that the main mechanism of their aggregation is the formation of polysaccharide slime connecting the cells together. There are also known soil aggregates, which are the clumps of soil particles, microbial slime, bacterial cells, and fungal hyphae. These aggregates are formed due to the frame binding by the hyphaes of fungi and slime production binding the particulates together. Microbial granule, called mycetoma or sclerotia, can be formed inside the body of human or animal. It is spherical or ellipsoidal aggregate of slow-growing bacterial cells or fungal mycelium in the infected body part. Probably, these spherical granules are formed due to microbial growth in dense tissue, which is pressing aggregate of microbial cells evenly from all directions. Hypothetically, it would be possible to find granular microbial aggregates in all viscous and stagnant aquatic environments, where the mechanical dispersion of microbial cells from aggregate is weak.

Aerobically Grown Microbial Granules

Formation of fungal spherical pellets is a common phenomena during cultivation of mycelial fungi or acinomycetes in shaking flask or stirred fermentor. This aggregation was an important process in industrial synthesis of antibiotics by mycelial fungi or acinomycetes and actively studied in 1960s. Current interest to aerobically grown microbial granules is related to their formation during the wastewater treatment in sequencing batch reactors (SBR). Aerobically grown microbial granules are actively investigated as bioagents for the biological treatment of wastewater. The main advantages of these microbial aggregates over conventional microbial flocs used in the wastewater treatment are short settling times and the ability to treat high strength wastewater (Morgenroth et al., 1997; Beun et al., 1999, 2000, 2002; Peng et al., 1999; Etterer and Wilderer, 2001; Tay et al., 2001a,b, 2003a,b; Moy et al., 2002; Toh et al., 2002; Zhu and Wilderer, 2003). It is assumed that there will be no need to construct and use secondary settling tanks, occupying huge area of wastewater treatment plant, in case when microbial granules will be used as the bioagents in wastewater treatment.

Structural Features of Aerobically Grown Microbial Granules

The features of aerobically grown microbial granules, which are used in wastewater treatment, are as follows:

1. Spherical or ellipsoidal shape; sometimes they can be elongated so that they are rod-like;
2. Size from 0.2 to 7 mm;
3. Filamentous, smooth, or skin-like surface, which is dominantly hydrophobic or hydrophilic;
4. Gel-like interior (matrix); sometimes it contains black matter or gas vesicule in central part of big dense granule;
5. Layers and microaggregates of specific microorganisms;
6. Channels and pores;
7. Inclusions of particulates.

Shape and Size of the Granules

Typical shape of the granules are sphere or ellipsoid (Fig. 6.1a). The roundness is evaluated by the ratio of the shortest and longest axis of ellipsoid. Probably, elongation of the granule depends on air upflow velocity in SBR. The granules can be significantly elongated at a high upflow velocity of 1.3 cm/s, and after several months of cultivation (Fig. 6.1b).

However, there may be the granules with irregular shape (Fig. 6.1c) or small granules merged together and producing grape-shaped aggregate (Fig. 6.1d). The size of aerobically grown microbial granules varied in a wide range, from 0.2 to 10 mm, and depended on the balance of biomass growth, production of cell binding exopolysaccharides, and cell detachment from the granule. The relationships between these processes are shown in Table 6.1.

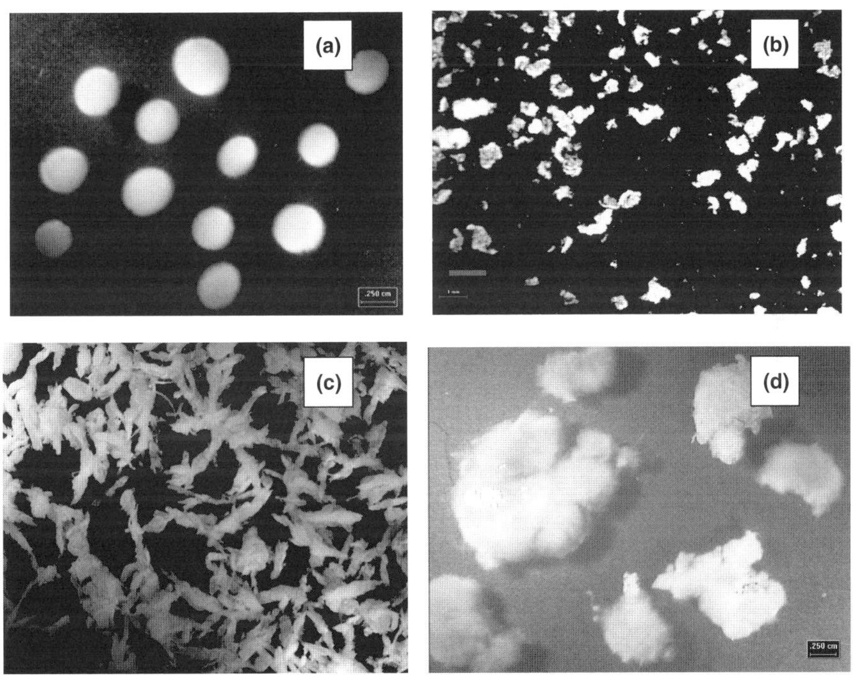

Fig. 6.1. Shape of the aerobically grown microbial granules (a) spherical and ellipsoid granules; (b) granules of irregular shape; (c) super-elongated granules produced at high upflow air velocity (photo from Dr. Liu Yongqiang); (d) granules produced by filamentous microorganisms (fungi, actinomycetes, filamentous bacteria). (See Color Plate Section before the Index.)

Table 6.1. Effect of cultivation conditions on granule size

Cell growth	Production of polysaccharides, which are bound cells in granule	Cell detachment from the granules due to aeration or stirring	Granule size*	Type of surface and its hydrophobicity
Fast growth due to rich medium and optimal conditions	No	Weak	Big	Filamentous hydrophilic or hydrophobic surface
Fast growth due to rich medium and optimal conditions	No	Strong	Medium	Smooth hydrophobic surface
Slow growth due to limitation by nitrogen and phosphorus (unbalanced catabolism and anabolism)	Yes	Weak	Medium	Smooth hydrophilic surface
Slow growth due to limitation by nitrogen and phosphorus (limitation of anabolism; unbalanced catabolism and anabolism)	Yes	Strong	Small	Smooth hydrophilic surface
Slow growth due to inhibition by toxic substances (inhibition of catabolism)	No	Weak or strong	Small	Smooth hydrophobic surface
Slow growth due to inhibition by toxic substances (inhibition of anabolism; unbalanced catabolism and anabolism)	Yes	Weak or strong	Small	Smooth hydrophilic surface

*The granules at stable period of cultivation in SBR with size less than 1 mm, between 1 and 5 mm, and bigger than 5 mm were considered conventionally as small, medium, and big ones, respectively.

Surface of Granules

The surface of the granules may be rough (filamentous) or smooth (Fig. 6.2) depending on the balance of biomass growth rate, production rate of cell binding exopolysaccharides, and cell detachment rate from the granule (Table 6.1). In some cases, cells of protozoa are attached to the surface of the granules (Fig. 6.2c,d). There are hydrophilic sites on the surface of the granules due to the presence of OH, COO^-, HPO_4^{2-}, NH_2, and other polarized groups of polysaccharides and proteins. Together with this, there may be a hydrophobic site caused by the presence of aliphatic chains and aromatic rings of lipids and proteins. The rule of the thumb is that the surface of the granule is dominantly hydrophilic if there is no production of exopolysaccharides in the granule (Table 6.1).

In cases with absence of excessive production of polysaccharides and strong aeration, the granules are covered by skin-like envelope, which is

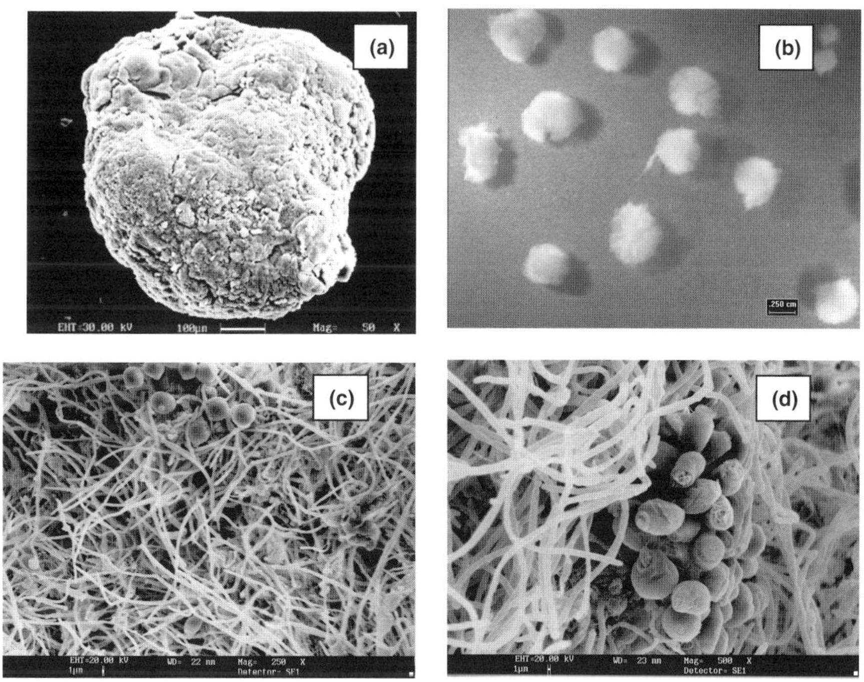

Fig. 6.2. Granules with smooth surface (a) and rough (filamentous) surface (b,c) and cells of protozoa attached to surface (d).

composed of dead cells. This skin-like envelope reduces cells detachment from the granule and protects it from mechanical destruction. If the pressure changes or mechanical impulses become too strong, the destruction of granule is due to crack in skin-like envelope of granule, and gel from the granule is released to the environment.

Radial Structures in Granule

Important structural property of microbial granules, related to their bioengineering functions, is arrangement of granule components as radial sub-aggregates, spherical sub-granules, and concentric layers (Fig. 6.3). Depth, thickness, and arrangement of these components can affect the formation, stability, and activity of the granules. The sub-aggregates inside the granules may be arranged randomly, in a radial direction (Fig. 6.3a), or as concentric layers (Fig. 6.3c). Larger granules may also result from merging of smaller granules.

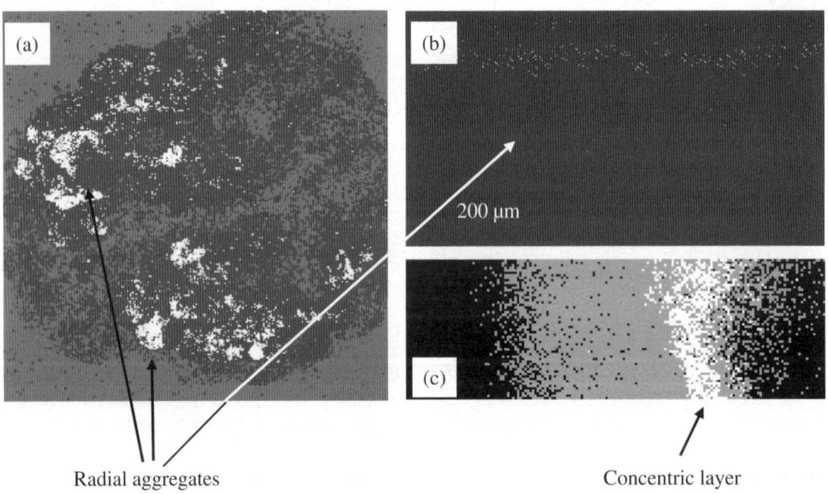

Fig. 6.3. Different microaggregates of cell aggregates (a) radially arranged microaggregates of ammonia-oxidizing bacteria (bright structures) in a 3D image produced by CLSM of the granule; (b) biofilm of nitrifying bacteria on the surface of Noble Agar with oxygen supply through the agar surface and ammonia supply from the agar bottom; one layer was a uniform biofilm but another one contained aggregates of nitrifying bacteria arranged perpendicular to the agar surface; (c) concentric layer of *Bacteroides spp*.

In spherical granular biofilm, ammonia-oxidizing bacteria were arranged in radially elongated aggregates within a layer with a depth from 70 to 100 μm from the surface of the biofilm (Tay et al. 2002a; Ivanov et al., 2005a,b) as shown in Figs 6.3a and 6.4b. Labeled oligonucleotide probes Bacto1080 and Nsm156 probes were applied using the FISH procedure.

Formation of radial aggregates of nitrifying bacteria (Figs 6.3a and 6.4b), in a direction that is normal to the granule surface, is probably driven by the co-existence of steep oxygen gradient and reverse ammonia gradient created by release of ammonia from the central core of granule where biomass is lyzed. The hypothesis on reverse gradient of ammonia in the granules was examined in independent experiments during the growth of enrichment culture of nitrifying bacteria in Noble Agar where oxygen was supplied through the agar surface but where ammonia was supplied from the bottom of the agar layer. Ammonia-oxidizing bacteria formed two layers. The first layer was a uniform biofilm but the second layer contained aggregates of nitrifying bacteria aligned normal to the agar surface (Fig. 6.3b). The decreasing widths of these nitrifying aggregates probably reflect the dependence of growth rate on the available concentration of dissolved oxygen and ammonia.

Cells of ammonia-oxidizing bacteria are often arranged in the laminar biofilms as microbial colonies embedded in slime attached to a carrier surface (Okabe et al., 1999). In laminar microbial biofilm on sea shells ammonia-oxidizing cells were arranged as a layer of vertically elongated aggregates (Ivanov et al., 2005b). These aggregates were embedded within the matrix formed by other bacteria. Vertically elongated aggregates seemed to be capable of multiplication due to their lateral growth and further splitting. Vertical (radial) cell aggregates may be ecologically important in bacterial biofilms because they have a higher surface-to-volume ratio (*S/V*) than laminated biofilms. For example, *S/V* for a 100 μm layer of biofilm is 0.01 $\mu m^2/\mu m^3$. However, if the microbial layer consists of vertical 20-μm-diameter cylinders, arranged so that the axes of neighboring cylinders are 40 μm apart, then $S/V = 0.21$. Therefore, when the microbial biofilm is arranged as a layer of vertical aggregates, the *S/V* ratio, and respectively, the rates of substrate transfer, microbial metabolism and growth, could be 20 times higher than the same parameters for laminated biofilms.

Vertically arranged, pear-shaped aggregates of ammonia-oxidizing bacteria have been shown in spherical suspended microbial biofilms

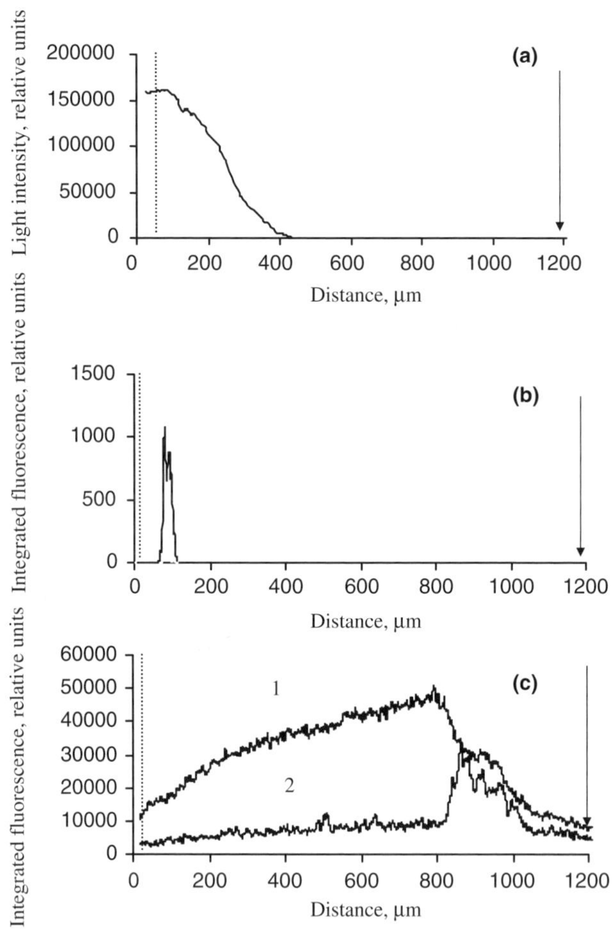

Fig. 6.4. Layers, channels, and pores in aerobic granules (a) shows an edge of the granule by evaluating the decrease of the light intensity through the granule; (b) shows a layer of aerobic, ammonia-oxidizing bacteria *Nitrosomonas spp*; (c) shows the distribution of microbial biomass (by measuring fluorescence intensity of hybridized Eub338 probe) (curve 1) and a layer of anaerobic bacteria *Bacteroides spp*. (curve 2); (d) shows the presence of the channels and the pores in the granule by measuring fluorescence of 0.1 μm microspheres penetrating into the granules from the medium; (e) shows the fluorescence of dead cells; (f) shows the fluorescence of polysaccharides stained by FITC-labeled ConA. The dotted line in all the figures represent the granule surface. The arrow in all the figures represent the granule center.

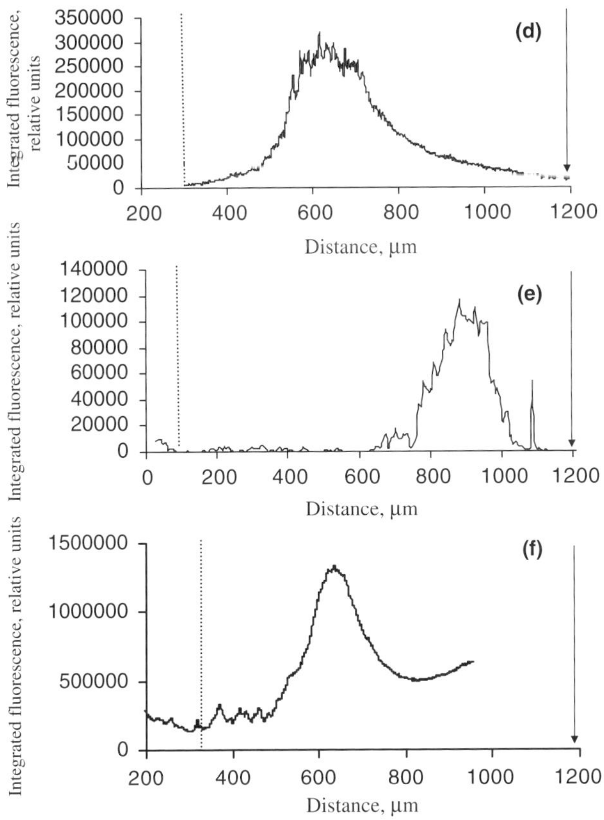

Fig. 6.4. — Cont'd.

(Ivanov et al., 2005b). It is likely that more examples of such vertically arranged aggregates in microbial biofilm could be found. Presence of these vertically arranged microbial aggregates must be taken into account in the models of microbial biofilms (Morgenroth et al., 2004; Picioreanu et al., 2004) and in the mathematical model of aerobic microbial granule.

Concentric Layers of Granule

The occurrence of concentric layers in granules was demonstrated using CLSM after staining by specific fluorochromes or FISH with specific oligonucleotide probes. The description of the layers is given in Table 6.2. Considering a microbial granule as a sphere with a diameter of 2.4 mm,

Table 6.2. Descriptions of layers in aerobically grown microbial granules grown in a column SBR with a medium containing ethanol or acetate (Tay et al., 2002a,b)

Layer	Average depth of layer from the surface of granule and average thickness	Assumed function in the granule
Aerobic ammonia-oxidizing bacteria	70 μm (depth); 30 μm (thickness)	It reflects the depth of oxygen diffusion into granule
Facultative anaerobic enterobacteria	Concentration increased to maximum at a depth of 450 μm and remained stable at depths from 450 to 850 μm	Bacteria perform both aerobic and anaerobic processes
Obligate anaerobic bacteria *Bacteroides spp.*	850 μm (depth); 150 μm (thickness)	It reflects the presence anaerobic zone in granule
Channels and pores by penetration of 0.1 μm microspheres	Depth linearly depends on granule diameter by equation (6.1)	Deeper diffusion of nutrients
Layer of active biomass	Thickness linearly depends on granule diameter	All bioactivities of the granule are concentrated in this layer
Polysaccharides	Low content to a depth of 500 μm, reaching a maximum at 650 μm. Stable but low content at depth from 800 to 1200 μm	It can decrease diffusion of nutrients into granule through the channels
Core of dying cells in the center of granule	Depth was 1000 μm. Diameter of this inner core depended on granule diameter	Supply of monomers and ammonia from this zone

the volumes of different zones can be calculated and compared with the experimental microbiological diversity of the granules (Table 6.3).

To determine the percentage of aerobic, facultative anaerobic, and anaerobic bacteria, cloning and sequencing of the 16S rRNA genes of the bacteria in the granules and phylogenetic analyses of the cloned sequences

Table 6.3. Average geometric and biological parameters of 2.4 mm spherical granule grown in a column SBR with a medium containing ethanol or acetate

Layer or zone in the granule	Geometric parameters	Volume, mm^3	% of total volume of the granule	% of related bacterial clones isolated from the granules
Aerobic microorganisms in porous layer	0.55 mm below granule surface	6.09	84.1	$69 \pm 7\%$
Facultative anaerobic microorganisms	Between 0.55 and 0.85 mm below granule surface	0.97	13.4	$9 \pm 7\%$
Obligate anaerobic microorganisms (*Bacteroides spp.*)	Between 0.85 and 1.0 mm	0.15	2.0	2.1%
Central core of dead and lyzed cells	Depth is 1 mm. Diameter is 0.4 mm	0.03	0.5	–

were performed (Tay et al., 2002b). Physiological property of the operational taxonomic units (OTUs), relation to oxygen, was inferred from the phylogenetic identification of OTUs. A CY5-labeled Ent1432 probe with the sequence 5′-CTTTTGCAACCCACT-3′ (Sghir et al., 2000) and with T_m of 45°C was used to detect enterobacteria. There is a statistically reliable correlation between the calculated volumes occupied by aerobic, facultative anaerobic, and anaerobic bacteria and the experimentally determined percentages of aerobic, facultative anaerobic, and anaerobic bacteria isolated from the granules.

Biomass and Polysaccharides in Granule

The exopolysaccharide (EPS) matrix in the granule was detected with a FITC-labeled lectin (ConA-FITC) from *Canavalia ensiformis*. The LIVE/DEAD® BacLight Bacterial Viability Kit (Molecular Probes, OR, USA) was used to evaluate quantity of dead and viable biomass. Intensity of green fluorescence of biomass stained with SYTO™ from this kit correlated with the number of ribosomes in cells, which depends on specific growth rate. Therefore, staining with SYTO™ was used for the detection of the active biomass layer in the granule.

The anaerobic layer in the aerobically grown granule was detected by the presence of obligate anaerobic bacteria *Bacteroides spp*. This layer was situated at a depth of 800–900 μm from the surface of the granule (Fig. 6.4c). The layer of anaerobic bacteria was followed by a layer of dead microbial cells at a depth of 800–1000 μm from the surface of the granule (Fig. 6.4e). Anaerobiosis and cell death in the granule interior was probably promoted by the formation of polysaccharide plugs in the channels and pores. These plugs diminished the mass transfer rate of both nutrients and metabolites. Polysaccharide formation peaked at a depth of 400 μm from the surface of the granule (Fig. 6.4f).

The core in some granules contains dead microbial cells and polysaccharides at a depth of 800–1000 μm from the surface of the granule (Fig. 6.4e). Cell death in the core was probably promoted by the formation of polysaccharide plugs in the channels and pores.

Channels and Pores

TetraSpec Fluorescent Microsphere Standards (Molecular Probes, OR, USA) detected channels and pores with diameters greater than 0.1 μm. All were visualized with Fluoview300 confocal laser scanning microscope (CLSM) (Olympus, Japan) as described previously (Tay et al., 2002a,b). Observations with CLSM at 1000× magnification showed that the beads did not adhere to the cell surface. Therefore, their distribution within the granule is not a measure of the adsorption of the beads onto the granule matrix, but indicates the penetration of the beads into the granule interior by passage through pore and channel structures which have to be larger than 0.1 μm in size. The incubation period of 4 h was more than sufficient to allow complete penetration of the beads into the granule interior. Test measurements performed using different incubation times showed that bead penetration reached saturation levels within 1 h of incubation.

Mass transfer rate in microbial aggregates may be enhanced by the formation of channels and pores that interconnect the surface and the interior. Such channels and pores had been previously observed in aerobic biofilms (Massol-Deya et al., 1995). The aerobic granules in this current study also contain channels and pores that penetrated to depths of up to 900 μm from the surface of the granule (Fig. 6.4d). Channels and pores were detected in the granule and porosity values peaked at depths of 300–500 μm from the granule surface (Fig. 6.4d). The thickness of the porous layer in the granule positively correlated with the granule diameter. For example, a granule with a diameter of 550 μm had a porous layer with a thickness of 250 μm, and a granule with a diameter of 1000 μm had a porous layer with a thickness of 350 μm. The biomass and the porosity profiles were also observed to drop at the same depth below the granule surface. There was no penetration of 0.1 μm microspheres to the central core of the large granules (Ivanov et al., 2004).

Adherence and Release of Cells and Particles

The deterioration of the granules was studied by labeling cells with 1 μg L^{-1} of fluorescent lipophilic tracer $DiIC_{18}(3)$ (Molecular Probes, OR, USA). The tracer was readily taken up by the cells. The in-solution concentration of the tracer one day after its introduction into the reactor was less than 1% of the concentration detected in a suspension of particles produced by disintegrating granules in a 2 mL tube with phosphate-buffered saline (PBS) using a Mini-Beadbeater (Biospec Products, Inc., Bartlesville, OK, USA) for 100 s at 500 rpm. The amount of fluorescence due to the lipophilic stain was determined using a Luminescence Spectrometer LS-50B (Perkin-Elmer, Boston, MA 02118, USA). Background due to autofluorescence of biomass was excluded from the reported values.

The granules are not only able to degrade organic matter but are also able to remove nano- and microparticles from wastewater due to microchannels and pores in the matrix of the granules. To detect the removal of 0.1 μm, 0.6 μm, 4.2 μm fluorescent microspheres, and cells of *Escherichia coli*, stained by permeable nucleic acid stain SYTO9™, the granules were incubated with these particles. Total number of the particles bigger than 0.1 μm in the reactors was approximately 4×10^7 per mL, and 23% of these particles were bacterial cells. The cells of *Escherichia coli* and 4.2 μm microbeads were accumulated within 250 μm in the upper layer of the microbial granule but small 0.1 μm microbeads penetrated to the depth approximately 500 μm in the granules (Fig. 6.5).

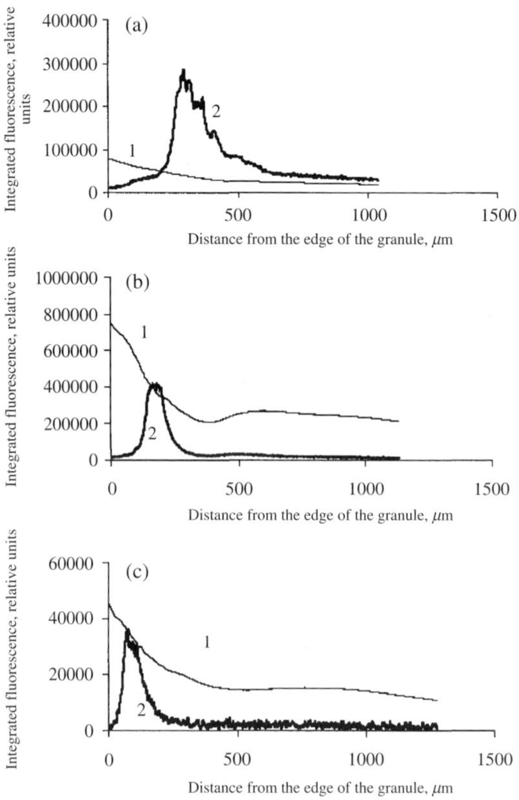

Fig. 6.5. Penetration of 0.1 μm beads (a), cells of *E. coli* (b), and 4.2 μm beads into the granule (c). Curve 1 shows the intensity of transparent light.

Microbial granules contained also attached ciliates (Fig. 6.2d) but accumulation of the particles in protozoan cells was smaller than in the granule matrix. Kinetics of particle sorption was revealed by flow cytometry and fluorescence spectrometry. Almost half of the stained cells of *E. coli* can be removed by the granules for one hour. The ability of the microbial granules to remove the particles can enhance their function in aerobic treatment of wastewater.

Anaerobic Processes in Aerobically Grown Granules

Due to the dense aggregation of cells, the rate of mass transfer of nutrients and metabolites between bulk medium and granular matrix may not be sufficient to ensure normal cell metabolism in the granule interior. The concentration of dissolved oxygen can drop to zero at some depth below the granule surface. This depth depends on the specific rate of oxygen consumption and also on the porosity and extent of channel structures in the granules. The typical depth of the aerobic zone in a thick microbial biofilm in the presence of aeration is between 50 and 200 μm (Villaverde and Fernandez-Polanco, 1999; Gieseke et al., 2001). The Bacto1080 probe with the sequence 5′-GCACTTAAGCCGACACCT-3′ is specific for *Bacteroides spp.* (Sghir et al., 2000) and was labeled and used to detect obligate cells of anaerobic *Bacteroides spp.* It was demonstrated that obligate anaerobic bacteria can grow in the interior of aerobically grown granules (Tay et al., 2002a,b, 2003a). Optical and mechanical sectioning of the granules following FISH incubation showed that the typical structure of a granule could be described as a sack consisting of thick walls of active biomass. In the granules with the walls approximately 1000 μm thick, the cells of the *Bacteroides* group were concentrated in a layer approximately 100 μm thick. This layer was located at a depth of 800–850 μm below the granule surface.

Optimization of Granule Size

The concentric layers were typically arranged in sequence as obligate aerobic bacteria, facultative anaerobic bacteria, obligate anaerobic bacteria, and finally a core of dead and lyzed cells. The presence of anaerobic bacteria can potentially diminish the stability of the granules due to the production of acids and gases from fermentation. Another negative effect of anaerobic bacteria on the wastewater treatment process is the occurrence of floating granules, which could occur if anaerobic bacteria are allowed to incubate in medium containing nitrate accumulated due to nitrification (Fig. 6.6). There were anaerobic conditions in the layer of settled granules. Therefore, floating of the granules was probably due to gas production during denitrification, similar to the floatation of denitirifying granules (Etchebehere et al., 2002). This potential floating of the microbial granules in case of high organic or nitrate load leading to the production of gases in anaerobic zone of the granule can deteriorate wastewater treatment.

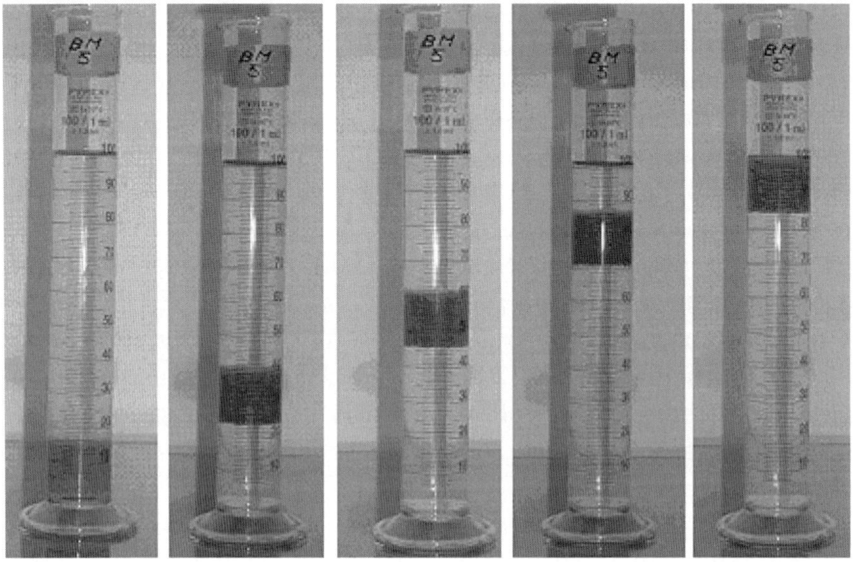

Fig. 6.6. Floating of the granules after settling. Fig. 6.6a to 6.6e correspond to 2, 24, 25, 28, and 32 min after settling of the granules. The sample was collected at the end of one cycle of cultivation in SBR fed with synthetic wastewater which was composed of 1000 mg L^{-1} of COD (ethanol), 300 mg L^{-1} of ammonia nitrogen, 2400 mg L^{-1} of bicarbonate, and micronutrients. Almost all ammonia was oxidized to nitrate by the end of cycle.

To avoid the formation of anaerobic layer and core and possible deterioration of wastewater treatment, the aerobic granules should have a diameter that is less than twice the distance from the granule surface to the anaerobic layer. This minimal distance is 850 μm (Table 6.2). Therefore, diameter of the granules without anaerobic layer and core of lyzed cells should be less than 1.7 mm.

Another approach of size optimization is based on the assumption that the entire granules should have a porous biomass-filled matrix without a core filled by dead and lyzed cells. Depth and thickness (H_l) of the layer of porous biomass linearly correlated with granule diameter (D_g) by equation (6.1):

$$H_l = 0.15\,\text{mm} + 0.2D_g \tag{6.1}$$

The optimal size of the aerobically grown microbial granule (D_c) may be calculated from equation (6.1) using the condition that $2H_l = D_g$, which means that whole granule with diameter less than or equal to D_c is consisting entirely of a porous matrix. The value of D_c calculated from equation (6.1) for this condition is 0.5 mm.

Physiological parameters such as specific COD removal or oxygen uptake rate cannot be used for conclusion on optimal diameter of the granules because increase of granule size diminishes the TOC and COD removal rate per 1g of VSS of the granules (Toh et al., 2002). The optimal diameter of the studied aerobic granule is less than 1.7 mm considering absence of the layer of obligate anaerobic bacteria or less than 0.5 mm considering that the whole granule should have a porous biomass-filled matrix. Design of the granulation process and reactor must include the condition to select or retain in the reactor the granules with a diameter smaller than the critical diameter. This critical diameter may be substrate- and process-specific parameter.

Dynamics of Granule Formation and Destruction

The granules were retained in the SBR while the flocs were washed out with the effluent. Concentration of granular biomass (VSS) during 6 days of experimental period was stable, at $6.5 \pm 0.2\,\mathrm{g\,L^{-1}}$. Concentration of floc biomass (VSS) was $0.15 \pm 0.02\,\mathrm{g\,L^{-1}}$. Stable concentration of granular biomass can be due to the balanced attachment and detachment of the flocs to granules or balanced growth and destruction of the granules. The hydraulic residence time was 0.33*d*, which corresponded to a daily exchange of three reactor volumes. Therefore, the ratio of produced granular biomass to produced floccular biomass was 14.5.

This ratio was close to 18.3, the initial ratio of granular labeled biomass to the flocculent labeled biomass after 4 h of labeling with lipophilic tracer (one growth cycle in SBR). Content of lipophilic tracer in granular biomass was stable for 6 days of study and was thought to be attributed to balanced attachment and detachment of the flocs to granules or balanced growth and destruction of the granules. It cannot be regarded as the result of negligible degradation of granules because the labeled biomass was permanently released as the labeled flocs. The tracer content could decrease if the rate of granule growth is higher than the rate of granule degradation.

References

Beun, J.J., Hendriks, A., van Loosdrecht, M.C.M., Morgenroth, E., Wilderer, P.A., & Heijnen, J.J. (1999). Aerobic granulation in a sequencing batch reactor. *Water Res.*, *33* (10), 2283–2290.

Beun, J.J., van Loosdrecht, M.C.M., & Heijnen, J.J. (2000). Aerobic granulation. *Water Sci. Technol.*, *41* (4–5), 41–48.

Beun, J.J., van Loosdrecht, M.C.M., & Heijnen, J.J. (2002). Aerobic granulation in a sequencing batch airlift reactor. *Water Res.*, *36* (3), 702–712.

Etchebehere, C., Errazquin, M.I., Cabezas, A., Pianzzola, M.J., Mallo, M., Lombardi, P., Ottonello, G., Borzacconi, L., & Muxí, L. (2002). Sludge bed development in denitrifying reactors using different inocula-performance and microbiological aspects. *Water Sci. Technol.*, *45* (10), 365–370.

Etterer, T., & Wilderer, P.A. (2001). Generation and properties of aerobic granular sludge. *Water Sci. Technol.*, *43* (3), 19–26.

Gieseke, A., Purkhold, U., Wagner, M., Amann, R., & Schramm, A. (2001). Community structure and activity dynamics of nitrifying bacteria in a phosphate-removing biofilm. *Appl. Environ. Microbiol.*, *67* (3), 1351–1362.

Ivanov, V., Tay, J.-H., Tay, S.T.-L., Tay, H.-L., & Jiang, R. (2004). Removal of micro-particles by microbial granules used for aerobic wastewater treatment. *Water Sci. Technol.*, *50* (12), 147–154.

Ivanov, V., Tay, S.T.-L., Liu, Q.-S., Wang, X.-H., Wang, Z-.W., & Tay, J.-H. (2005a). Formation and structure of granulated microbial aggregates used in aerobic wastewater treatment. *Water Sci. Technol.*, *52* (7), 13–19.

Ivanov, V., Tay, J.-H., Liu, Q.-S., Wang, X.-H., Wang, Z-.W, Maszenan, A.M., Yi, S., Zhuang, W.-Q., Liu, Y.-Q., Pan, S., & Tay, S.T.-L. (2005b). Microstructural optimization of wastewater treatment by aerobic granular sludge. *Aerobic Granular Sludge* (eds. Bathe, S., de Kreuk, M.K., McSwain, B.S., & Schwarzenbeck, N.), Water and Environmental Management Series. IWA Publishing, London, 43–52.

Liu, Y.-Q., Tay, J.W., Ivanov, V., Moy, Y.-P.B., Yu, L., & Tay, T.-L.S. (2005). Influence of phenol on nitrification by microbial granules. *Process Biochem.*, *40* (10), 3285–3289.

Massol-Deya, A.A., Whallon, J., Hickey, R.F., & Tiedje, J.M. (1995). Channel structures in aerobic biofilms of fixed-film reactors treating contaminated groundwater. *Appl. Environ. Microbiol.*, *61* (2), 769–777.

Morgenroth, E., Sherden, T., van Loosdrecht, M.C.M., Heijnen, J.J., & Wilderer, P.A. (1997). Aerobic granule sludge in a sequencing batch reactor. *Water Res.*, *31* (12), 3191–3194.

Morgenroth, E., Eberl, H.J., van Loosdrecht, M.C.M., Noguera, D.R., Pizarro, G.E., Picioreanu, C., Rittmann, B.E., Schwarz, A.O., & Wanner, O. (2004). Comparing biofilm models for a single species biofilm system. *Water Sci. Technol.*, *49*, 145–154.

Moy, B.Y.P., Tay, J.H., Toh, S.K., Liu, Y., & Tay, S.T.L. (2002). High organic loading influences the physical characteristics of aerobic sludge granules. *Lett. Appl. Microbiol.*, *34*, 407–412.

Okabe, S., Satoh, H., & Watanabe, Y. (1999). *In situ* analysis of nitrifying biofilms as determined by *in situ* hybridization and the use of microelectrodes. *Appl. Environ. Microbiol.*, *65* (7), 3182–3191.

Peng, D.C., Bernet, N., Delgenes, J.-P., & Moletta, R. (1999). Aerobic granular sludge – a case report. *Water Res.*, *33* (3), 890–893.

Picioreanu, C., Kreft, J.-U., & van Loosdrecht, M.C.M. (2004). Particle-based multidimensional multispecies biofilm model. *Appl. Environ. Microbiol.*, *70*, 3024–3040.

Sghir, A., Gramet, G., Suau, A., Rochet, V., Pochart, P., & Dore, J. (2000). Quantification of bacterial groups within human fecal flora by oligonucleotide probe hybridization. *Appl. Environ. Microbiol.*, *66* (5), 2263–2266.

Tay, J.H., Liu, Q.S., & Liu, Y. (2001a). Microscopic observation of aerobic granulation in sequential aerobic sludge reactor. *J. Appl. Microbiol.*, *91* (1), 168–175.

Tay, J.-H., Liu, Q.-S., & Liu, Y. (2001b). The role of cellular polysaccharides in the formation and stability of aerobic granules. *Lett. Appl. Microbiol.*, *33* (3), 222–227.

Tay, J.-H., Ivanov, V., Pan, S., & Tay, S.T.-L. (2002a). Specific layers in aerobically grown microbial granules. *Lett. Appl. Microbiol.*, *34* (4), 254–258.

Tay, S.T-L., Ivanov, V., Yi, S., Zhuang, W.-Q., & Tay, J.-H. (2002b). Presence of anaerobic *Bacteroides* in aerobically grown microbial granules. *Microb. Ecol.*, *44* (3), 278–285.

Tay, J.-H., Tay, S.T.-L., Ivanov, V., Pan, S., Jiang, H.-L., & Liu, Q.-S. (2003a). Distribution of biomass and porosity profiles in microbial granules used for aerobic wastewater treatment. *Lett. Appl. Microbiol.*, *36* (5), 297–301.

Tay, J.-H., Pan, S., Tay, S.T.-L., Ivanov, V., & Liu, Y. (2003b). The effect of organic loading rate on the aerobic granulation: the development of shear force theory. *Water Sci. Technol.*, *47* (11), 235–240.

Tay, J.-H., Tay, S.T.-L., Ivanov, V., & Hung, Y.-T. (2004). Application of biotechnology for industrial waste treatment. *Handbook of Industrial Wastes Treatment*. 2nd edition, revised and expanded (eds. Lawrence K. Wang, Yung-Tse Hung, Howard H. Lo, & Constantine Yapijakis), Marcel Dekker, New York, 585–618.

Toh, S.K., Tay, J.H., Moy, B.Y.P., Ivanov, V., & Tay, S.T.-L. (2002). Size effect on the physical characteristics of the aerobic granule in a SBR. *Appl. Microbiol. Biotechnol.*, *60* (6), 687–695.

Villaverde, S., & Fernandez-Polanco, F. (1999). Spatial distribution of respiratory activity in *Pseudomonas putida* 54G biofilms degrading volatile organic compounds (VOC). *Appl. Microbiol. Biotechnol.*, *51* (3), 382–387.

Zhu, J.-R., & Wilderer, P.A. (2003). Effect of extended idle conditions on structure and activity of granular activated sludge. *Water Res.*, *37* (9), 2013–2018.

Chapter 7

Microorganisms of Aerobic Microbial Granules

Volodymyr Ivanov and Stephen Tiong-Lee Tay

Granules as Cellular Aggregates

A multicellular aggregate is formed and separated from its surrounding environment due to:

(1) Aggregation by hydrophobic force, electrostatic interactions, or salt bridges;
(2) Loose polysaccharide or inorganic matrix (iron hydroxide as example) combining the cells together by mechanical embedding, chemical bonds, hydrogen bonds, electrostatic forces, or hydrophobic interactions;
(3) Formation of mycelia, which is a net of branched cell filaments;
(4) Polysaccharide matrix with a filamentous frame;
(5) Structured matrix with layers parallel to the boundary or sub-aggregates, which are perpendicular to the boundary;
(6) Coverage by a common sheath of organic (polysaccharides, proteins) or inorganic origin (iron hydroxide, silica, calcium carbonate);
(7) Coverage by a common sheath ("skin" of microbial aggregate) consisting of dead cells.

Aerobically grown microbial granules are aggregates which are specified as follows:

(1) with regular shape (spherical, egg-shaped, or elongated oval in cross section);
(2) with size from 0.5 to several mm;
(3) with high-settling velocity from 0.2 to 2 cm/s;
(4) with high density and sludge-to-volume index (SVI) from 20 to 80 g/l.

Microbial Interactions in Aggregates

The population density or average distance between microbial cells determines the type of interaction (Fig. 7.1).

Our general view is that when the population density is low, organisms have neither positive, nor negative, interactions. When the population density is medium, organisms compete among themselves for the availability of resources, by rate or efficiency of growth, and by production of metabolites, which negatively affect the growth of competitors. When the population density is high, cells usually aggregate and cooperate between themselves. Therefore, cooperation, not competition, can be expected between microorganisms isolated from aerobically grown microbial granules.

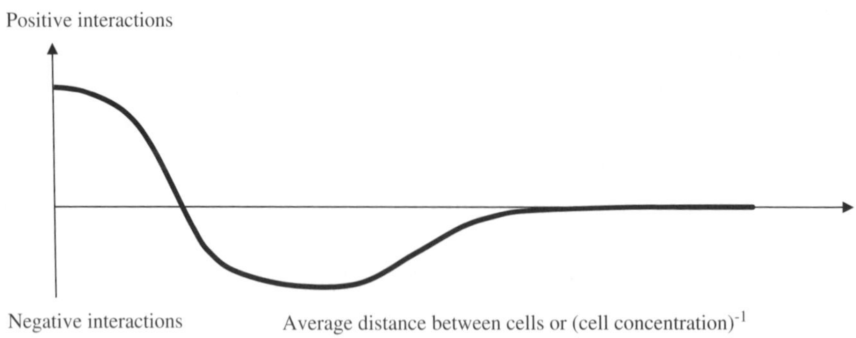

Fig. 7.1. Microbial interactions depending on cell concentration in ecosystem or the distance between cells in community.

Both competition and cooperation are carried out mainly due to the changes of chemical factors of environment such as concentration of nutrients, pH, and redox potential of the medium, excretion of antibiotics, extracellular digestive enzymes, or heavy metals binding exopolysaccharides, simultaneous biodegradation of substances.

Hypothetically, cooperation between microorganisms in the granules might be aimed for enhancement of the following functions:

(1) cell aggregation;
(2) formation of flexible mechanical frame of the granule by filamentous microorganisms;
(3) sequential utilization of carbon source, especially xenobiotics;
(4) formation of the intragranular storages of carbon (extracellular and intracellular storage of polysaccharides), phosphate (intracellular polyphosphate), iron, and other nutrients;
(5) formation of sheath or envelope of granule, protecting cells in the interior from toxic substances or unfavorable environmental conditions.

For example, there might be following mechanisms of commensalisms or mutualism between the microorganisms in the granule:

(1) facultative anaerobes use oxygen and create the conditions for the growth of obligate anaerobes inside granule; this interaction is important in the formation of anaerobic layer of granule grown under aerobic conditions;
(2) one strain produces a growth factor essential for another strain; this interaction is especially important in granules grown in simple medium with sole source of carbon;
(3) biodegradation of xenobiotic is performed as a sequence biochemical reactions by different microbial strains; in this case total rate of biodegradation will depend on the activities of different microbial groups in the granule.

A microbial aggregate can be considered as a multicellular organism if its parts have different coordinated or synchronized physiological functions, i.e. growth, motility, sexual interactions, assimilation of atmospheric nitrogen, production of extracellular polysaccharides, transport and distribution of nutrients, and reduction of oxygen.

Study of Microbial Community Diversity

Even hypothetical considerations of cooperation in the granule demonstrate that there might be diverse microbial community containing microorganisms with different physiological functions. Additionally, the heterogeneity of microbial community may be created by spatial diversity of environmental conditions in the granule due to existence of different zones, layers, aggregates, and chemical or physical gradients in the granule. Another aspect of microbial diversity is temporal changes of diversity as a succession from flocs to pro-granules and then to young granules, following by stagnation or climax, characterized by weak temporal changes caused by degeneration of matured microbial granules.

The analysis of microbial community residing in the aerobically grown granule can provide information on the microorganisms responsible for granule formation, maintenance, and activity. This knowledge can be used to better control of aerobic granulation.

Microbial populations present in wastewater treatment plants have been studied conventionally by culturing bacterial isolates (Snaidr et al., 1997). These culture-dependent methods suffer from several limitations (Moyer et al., 1994; Amann et al., 1995; Head et al., 1998) and therefore are inadequate to represent the *in situ* diversity and ecophysiology for a meaningful analysis of community structure or specific organism functionality.

All microorganisms can be detected also by culture-independent methods (Amann et al., 1995, 1998). These methods usually involve the identification of 16S rRNA genes using DNA extraction from the environmental sample, then amplification of gene sequence using polymerase chain reaction (PCR), cloning or separation of the sequences and determination of nucleotide sequence (Head et al., 1998). Culture-independent techniques based on ribosomal ribonucleic acid (RNA) provide a more comprehensive, rapid, and concise characterization of bacteria taxa present in discrete habitats.

Microbial Diversity Studied by Cloning–Sequencing Method

One culture-independent method is the PCR-cloning method which includes the following steps: (1) isolating native DNA from the natural community; (2) PCR amplification of small subunit (SSU) rRNA gene sequences with universal primers; (3) screening of clones for genetic variability; and (4) using these detected variations to estimate genetic diversity and to select clones for subsequent sequencing to determine phylogenic affiliation. For screening SSU rDNA clone libraries, the amplified ribosomal DNA restriction analysis (ARDRA) of 16S rRNA genes has become an effective strategy to identify putative operational taxonomic units (OTUs) in various microbial communities such as those from hydrothermal vent system (Moyer et al., 1994); subsurface soil (Chandler et al., 1997); marine sediment (Urakawa et al., 1999); contaminated aquifer (Dojka et al., 1998); and also in activated sludge wastewater treatment plants (Gich et al., 2000).

In general, ARDRA is an effective strategy to provide a general view of how the clone libraries differ and to identify important OTUs for further analysis (Moyer et al., 1994; Chandler et al., 1997; Urakawa et al., 1999; Gich et al., 2000). Estimating community structure and diversity at the DNA level is an invaluable tool for microbial ecology, but this strategy also has its potential problems and limitations (von Wintzingerode et al., 1997; Head et al., 1998). Although the cloning–sequencing method requires considerable time and effort, it can provide new insights and a better understanding of the phylogenetic diversity of different microbial habitats.

Various diversity indices can be used to compare the bacterial communities associated with the three clone libraries. Species richness, which represents the total number of species or operational taxonomic units (OTUs), was calculated by rarefaction (Cho and Kim, 2000) with the online Rarefaction Calculator (http://gause.biology.ualberta.ca/jbrzusto/rarefact.html). Bacterial diversity was calculated on the basis of RFLP types by using the Shannon–Weaver index (H), Pielou's evenness index (e), Simpson's dominance index (c), and equitability (J). The estimated percentages of coverage (Dang and Lovell, 2000) for the different libraries were calculated as follows: $[1-(n/N)] \times 100$, where n is the number of unique clones detected in a subsample (library) of size N.

Growth Stages of Aerobic Granules

Aerobic granules were obtained from a laboratory-scale sequential aerobic sludge blanket (SASB) bioreactor fed with glucose as the main source of

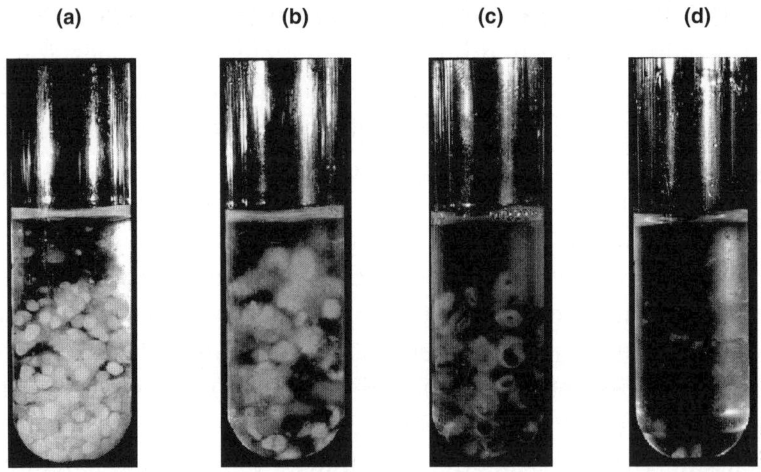

Fig. 7.2. Four stages of aerobic granules development: (a) young granules; (b) mature granules; (c) old granules with black cores; and (d) disintegrated granules. (See Color Plate Section before the Index.)

carbon and energy. Aerobic granules at different growth stages which are conventionally called as young, mature, old, and disrupted granules (Fig. 7.2), were selected manually using following principles:

(1) young granules are those of regular shape, smooth surface, and their diameters are smaller than 2 mm;
(2) mature granules are those with diameters bigger than 2 mm, fluffy edge, and without black core inside;
(3) old granules are those with diameters bigger than 1 mm and with black core inside;
(4) disrupted granules have irregular shape and appear as fluffy flocs.

The observation of the different growth stages of aerobic granules implied that the development of aerobic granules might go through a sequence of specific, but poorly understood chemical and biological processes. The investigation on tracking the microbial population dynamics from young to old granules would provide important information to identify the key organisms in aerobic granules development. This information would in turn provide useful physiological indicators for the purpose of monitoring and controlling this novel wastewater treatment process.

Amplified Ribosomal DNA Restriction Analysis (ARDRA)

The amplified 16S rDNA, full-length inserts (≈1500 bp), were selected for restriction enzyme digestion to define restriction fragment length polymorphism (RFLP) patterns using restriction endonuclease *Cfo*I (Promega, USA). To determine the similarities of bacterial populations present in three clone libraries, comparisons of the RFLP types within and among each of clone libraries were performed using GelCompar II (version 1.5) software. The constituent populations in the aerobic granules were defined in terms of operational taxonomic units (OTUs). The relative abundance of individual clones within each OTU was also assessed.

There was considerable diversity in each clone library and there were variations in microbial diversity among the three different clone libraries. This suggests a shift in the composition of the microbial communities. Microorganisms associated with five restriction fragment length polymorphism (RFLP) types (A, B, C, D, and E) appeared to play an important role in the development of aerobic granules.

A total of 144 clones containing the full-length inserts (≈1500 bp), were digested with restriction enzyme *Cfo*I, which has been shown to be particularly effective at defining operational taxonomic units (OTUs) (Moyer et al., 1996; Chandler et al., 1997). *Cfo*I digestion of full-length inserts resulted in 2–6 easily resolved bands, which were used for RFLP cluster analysis (data not shown). The results of 16S rDNA clone library and ARDRA are summarized in Table 7.1.

A total of 56 different RFLP types (OTUs), were identified from the three clone libraries. 21 RFLP types were detected from 45 clones derived from young granules, 27 types were identified from 52 clones from mature granules, and 23 types were recognized from 47 clones from old granules.

Table 7.1. Summary of 16S rDNA clones and RFLP types recovered from aerobic granule samples

Growth stages	Full-length clones	RFLP types	Unique clones	Coverage %
Young	45	21	11	75.6
Mature	52	27	17	67.3
Old	47	23	11	76.6
Total	144	56		

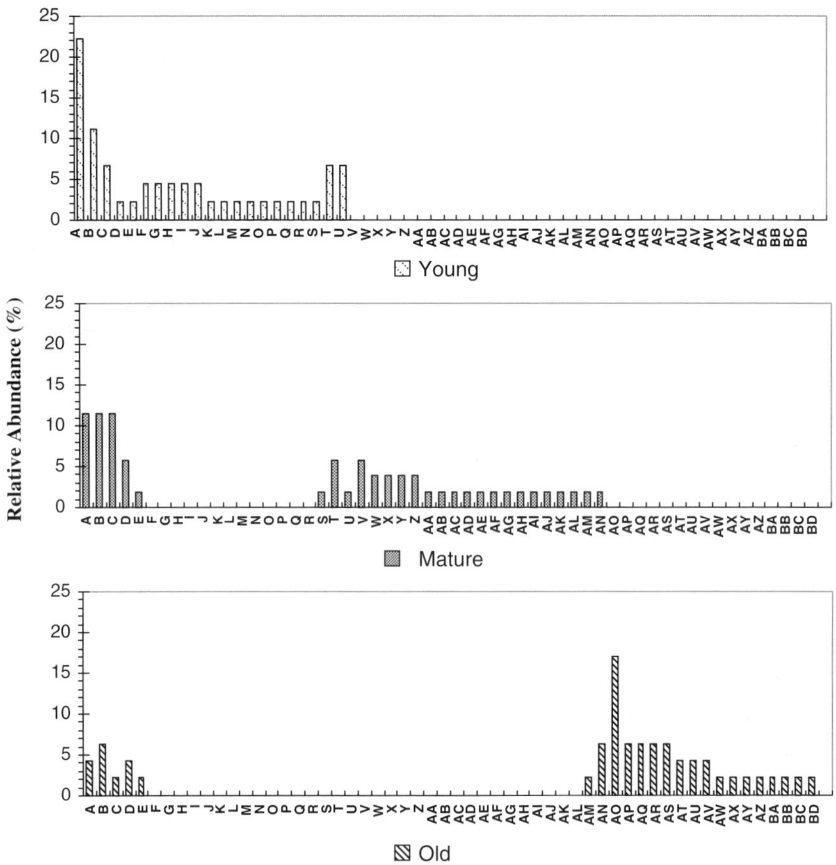

Fig. 7.3. Distribution and relative abundance of 16S rDNA clones from different growth stages of aerobic granules.

The distribution of clones in each clone library, both in RFLP types and relative abundance is plotted in Fig. 7.3.

Diversity Indices

The number of RFLP types (richness) and the frequency distribution of the RFLP types (evenness) in each of the clone libraries were evaluated by using a variety of standard diversity indices and results were summarized in Table 7.2.

Table 7.2. Diversity indices based on *Cfo*I RFLP patterns in 16S rDNA clone libraries from young, mature, and old granules

Parameter	Young	Mature	Old
RFLP type richness $E(S)$[a]	19.7	22.8	21.2
Shannon–Weaver diversity (H)[b]	2.742395	3.034237	2.927977
Evenness (E)[c]	0.900764	0.920627	0.933817
Simpson's dominance (c)[d]	0.09037	0.06213	0.067451
Equitability (J)[e]	0.72042	0.767919	0.760484

[a] $E(S)$ (Cho and Kim, 2000) was calculated by rarefaction for a standardized sample size of 40 clones for each clone library.
[b] H was calculated as follows: $H = -(p_i)(\ln p_i)$, where p_i is the proportion for each RFLP pattern.
[c] Evenness (E) was calculated from H as follows: $E = H/\ln S$, where S is the total number of RFLP patterns in each clone library.
[d] c was calculated as follows: $c = (p_i)^2$.
[e] J was calculated as follows: $J = H/H_{max}$, where $H_{max} = \mathrm{Log}_2 X$, where X is the total number of clones in each library.

Since the libraries differed in size, estimated RFLP type richness [$E(S)$] was calculated by rarefaction for smaller sample sizes (40 clones) to allow standardized comparisons. The estimated value of richness in the mature clone library was much higher than that in the young and old libraries. With the exception of Simpson's dominance index as shown in Table 7.2, the mature and old clone libraries had higher values on diversity indices such as richness, the Shannon–Weaver diversity index, evenness, and equitability than the young clone libraries. The young clone library contained a few RFLP types. Overall, the calculated diversity index values showed that the bacterial communities of mature and old granules were more diverse than that of the young granules. The values of coverage (Table 7.2) were also used to approximate the probability that all species present in a given sample were represented at least once in the library (Dang and Lovell, 2000). These three libraries had percent coverage values ranging from 68.6 to 77.1, which indicated that the microbial communities present in this study contained substantial diversity.

Microbial Community Analysis

Fifty-six RFLP types (OTUs) were identified based on the results of cluster analysis. Diversity indices revealed that bacterial communities of mature and old granules were more diverse than in young granules. Shifts in microbial populations were also confirmed by ARDRA. Such changes were attributed to physiological adaptation of bacteria during aerobic granulation process. Microorganisms associated with 5 OTUs (A, B, C, D, and E) appeared in all three clone libraries at different growth stages, which suggests that these bacteria may have an important role in the development of aerobic granules.

Shifts in bacterial community were observed in three clone libraries (Fig. 7.3). For example, 46 RFLP types detected were unique to each clone library (13 in young granules, 17 in mature granules, and 16 in old granules). Different RFLP types were also observed to dominate different clone libraries. In young granules, RFLP types A, B, C, T, and U were the most frequently occurring, but were not uniformly abundant throughout the different growth stages. In mature granules, the most commonly occurring RFLP types were A, B, C, D, T, and V. RFLP types B, AN, AO, AP, AQ, and AR dominated the old granules.

Several RFLP types were also found in more than one clone library. Eight RFLP types (A, B, C, D, E, S, T, and U) from the young granules were also found in the mature granules. Seven types (A, B, C, D, E, AM, and AN) from the mature granules were also found in the old granules. Five RFLP types, A, B, C, D, and E, appeared in all three libraries, which suggest that these 5 RFLP types may have important roles in the development of aerobic granules. The relative abundance of RFLP type A decreased from young to mature to old granules. On the other hand, the relative abundance of RFLP types B, C, and D increased slightly from young to mature granules and decreased significantly in old granules. The relative abundance of RFLP type E did not change from young to mature to old granules. This finding is important since the changes in relative abundance may reflect the onset of granule lysis.

As a consequence, the clone libraries should not be viewed as quantitative representations of microbial abundance in the original community. Nevertheless, some researchers (Farrelly et al., 1995; Suzuki et al., 1996) have demonstrated that changes on the composition of clone libraries can signal temporal (or spatial) variations within identical environmental matrices, and therefore should represent qualitative changes in the microbial community. Results obtained in this study showed significant differences in clone abundance at different growth stages of aerobic granules, which probably reflected relative changes in abundance of that gene or organism in the original community.

In this study, ARDRA was chosen as an initial measure of genetic diversity within each clone library and the response of the microbial community during granule development. Community changes in bacterial composition and the relative abundance would be due to interactions among different groups of bacteria and the microniches in which they reside. Community changes are therefore a consequence of the natural phenomenon of physiological adaptation by bacteria to the surrounding environment and their mutual interactions.

Aerobes and Facultative Anaerobes in Granules

Relation of detected OTUs to oxygen was inferred from their phylogenetic identification (Table 7.3).

Table 7.3. Distribution of OTUs by relation to oxygen

OTU name	% of clones found in the following library			Grouped with	Relation to oxygen, gliding motility in the closest species
	GA	GB	GC		
GB15	2.2	2.1	0	*Acidovorax sp.*	Aerobe
GC25	0	0	4.3	*Burkholderia cepacia*	Aerobe
GB38	0	4.2	0	*Comamonas acidovorans*	Aerobe
GA32	2.2	0	0	*Comamonas testosteroni*	Aerobe
GB7	0	2.1	0	*Janthinobacterium lividum*	Aerobe
GB42	0	4.2	0	*Herbaspirillum sp.*	Aerobe
GC45	0	2.1	4.3	*Janthinobacterium lividum*	Aerobe
GA16	2.2	0	0	*Janthinobacterium lividum*	Aerobe
GB40	15.6	33.3	4.3	*Janthinobacterium lividum*	Aerobe
GC7	0	0	6.4	*Janthinobacterium lividum*	Aerobe
GC23	0	0	2.1	*Frateuria aurantia*	Aerobe

(*continued*)

Table 7.3. Distribution of OTUs by relation to oxygen — Cont'd

OTU name	% of clones found in the following library			Grouped with	Relation to oxygen, gliding motility in the closest species
	GA	GB	GC		
GC53	0	0	17.0	*Pseudomonas fluorescence*	Aerobe
GA21	6.7	12.5	4.3	*Acidovorax sp.*	Aerobe
GC14	0	0	4.3	*Comamonas terrigena*	Aerobe
GA18	8.9	4.2	0	*Flavobacterium sp.*	Aerobe
GA23	4.4	0	0	*Flavobacterium sp.*	Aerobe
GA44	2.2	0	0	*Flavobacterium sp.*	Aerobe
GA36	4.4	0	0	*Flavobacterium balustinum*	Aerobe
GC11	0	0	12.7	*Flexibacter sp.*	Aerobe
GA53	2.2	0	0	*Cytophaga sp.*	Aerobe
GC52	15.6	10.4	4.3	*Cytophaga sp.*	Aerobe
Total aerobes	65.6	75.8	64.0		
GA55	11.1	0	0	*Erwinia persincina*	Facultative anaerobe
GC30	4.4	2.1	6.4	*Escherichia coli*	Facultative anaerobe
GC3	0	0	4.3	*Dysgonomonas sp.*	Facultative anaerobe
Total facultative anaerobes	15.5	2.1	10.7		
GB58	6.7	4.2	0	*Dechloromonas agitatus*	Anaerobe
GC40	0	2.1	2.1	*Lactococcus lactis*	Anaerobe
GB23	0	2.1	0	*Leuconostoc lactis*	Anaerobe
GC54	11.1	12.5	12.8	*Leuconostoc lactis*	Anaerobe
GC50	0	0	6.4	*Propionibacterium sp.*	Anaerobe
GC6	0	0	2.1	*Bacteroides sp.*	Anaerobe
GC51	0	0	2.1	*Bacteroides sp.*	Anaerobe
Total anaerobes	17.8	20.9	25.5		

Mean frequencies ±SD of OTUs related to obligate aerobes, facultative anaerobes, and anaerobes in all clone libraries were 69 ± 7%, 9 ± 7%, and 21 ± 4%. It was a statistically reliable correlation between percentage of total volume of the granule occupied by aerobic, facultative anaerobic, and obligate anaerobic bacteria and percentage of aerobic, facultative anaerobic, and obligate anaerobic bacterial clones isolated from

Table 7.4. Average geometric and biological parameters of 2.4 mm spherical granule grown in a column sequencing batch reactor with a medium containing ethanol or acetate

Layer or zone in the granule	Geometric parameters	Volume, mm^3	% of total volume of the granule	% of related bacterial clones isolated from the granules
Aerobic microorganisms in porous layer	0.55 mm below granule surface	6.09	84.1	69 ± 7%
Facultative anaerobic microorganisms	Between 0.55 and 0.85 mm below granule surface	0.97	13.4	9 ± 7%
Obligate anaerobic microorganisms (*Bacteroides spp.*)	Between 0.85 and 1.0 mm	0.15	2.0	2.1%
Central core of dead and lyzed cells	Depth is 1 mm. Diameter is 0.4 mm	0.03	0.5	

the granules (Table 7.4). The reason may be that cell concentrations of the representatives of OTUs in the granules were similar or at least with some narrow ranges.

Obligate aerobic ammonia-oxidizing bacteria were detected using FISH with oligonucleotide probes and CLSM within a layer with a depth from 70 to 100 μm from the surface of biofilm. Probably, this depth is related to the depth of oxygen diffusion into granule. At the same time, facultative anaerobic enterobacteria performing either aerobic or anaerobic processes were detected using FISH with oligonucleotide probes and CLSM within a layer with a depth from surface to 650 μm from the surface of granule. Concentration increased to maximum at a depth of 450 μm and remained almost stable at depths from 450 to 650 μm (Fig. 7.4).

Obligate Anaerobes in Granules

The representatives of phylogenetic groups in aerobically grown granules were determined and then hypothetically inferred physiological groups

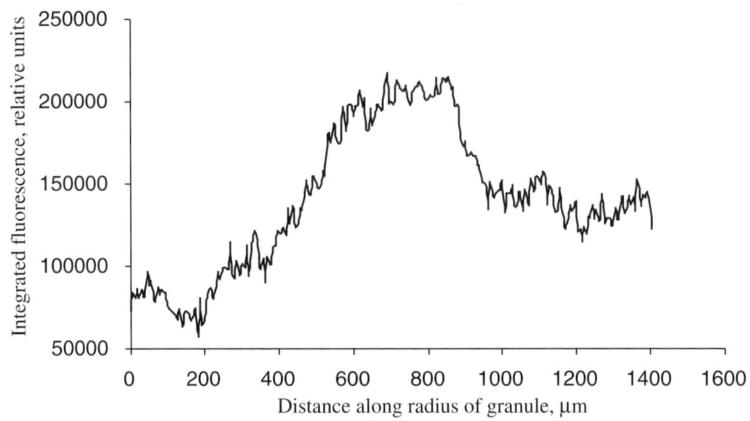

Fig. 7.4. Distribution of enterobacteria in aerobically grown granule.

of microbial community in granules have been used for analysis of microbial physiological diversity in the granules. One group includes obligate anaerobic bacteria from genus Bacteroides. Eleven clone sequence types were assigned to the *Cytophaga-Flavobacterium-Bacteroides* (CFB) group in clone library. A maximum likelihood tree generated by fast-DNAml program (Olsen et al., 1994) is shown in Fig. 7.5.

Clone 051 was affiliated with *Bacteroides fragilis*. A sequence of 06 clone was identified as belonging to *Bacteroides spp.* by its position on 16S rDNA phylogenetic tree (Fig. 7.5). It was closed to *B. distasonis* and *B. merdae*. Cloning of 16S rDNA and its sequencing demonstrated the presence of obligate anaerobe *Bacteroides spp.* in aerobically grown microbial granules.

These anaerobic bacteria were selected to detect the boundary of anaerobic microzone in the granules. Cells of *Bacteroides spp.*, detected by FISH and CLSM, were concentrated in a layer with a thickness about 100 μm. This layer was on the depth approximately 800 μm from the surface of the big granules. Cells of *Bacteroides spp.* were concentrated on inner surface of the walls of the big granule and were almost absent in the small granules.

Cells of *Bacteroides spp.*, which were stained after the incubation in hybridization buffer with Bacto1080 probe (Dore et al., 1998;

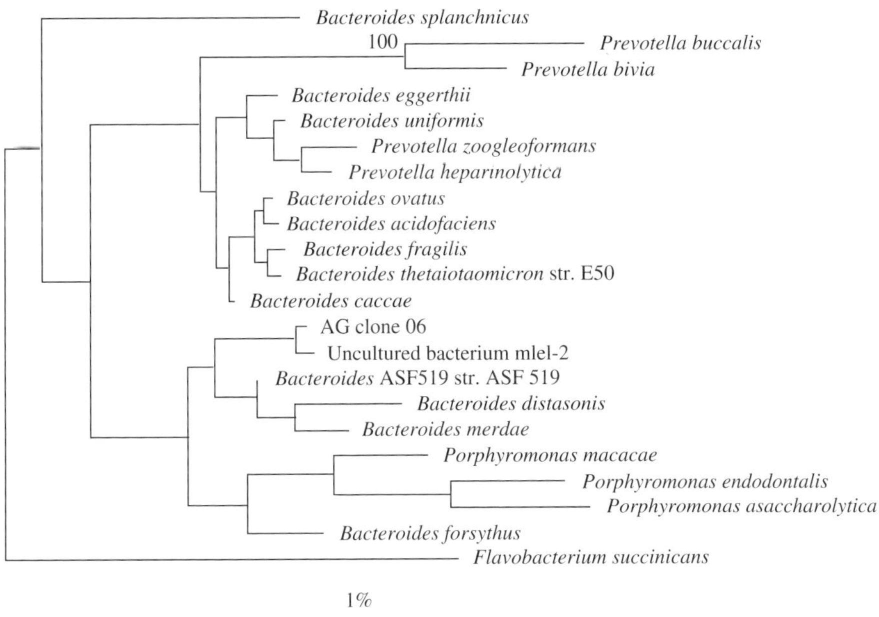

Fig. 7.5. 16S rRNA phylogenetic tree for cloned sequence 06. The numbers at the branch nodes are bootstrap values based on 100 resamplings for maximum likelihood. Only bootstrap values greater than 50% are shown. Scale bar = 1% nucleotide divergence.

Sghir et al., 2000) for FISH and washed in the washing buffer, were also clearly distinguished by flow cytometry on the dot plots showing red fluorescence (FL3) versus forward light scatter (FSC) (Fig. 7.6). The stained (hybridized) cells were concentrated in the region R2 and non-stained cells were appeared in the region R1 (Fig. 7.6b). These non-stained cells in the region R1 were determined in the control where the cells were treated as in the experiment but the Bacto1080 probe for detection of *Bacteroides spp.* was not added (Fig. 7.6a).

The ratio of the cells of *Bacteroides spp.* cells was determined as the ratio of the number of the events in the region R2 to the number of the events in the region R1. It was 0.56% for the cells from big granules and 0.22% for the cells from small granules. It is not absolute but a conventional ratio, because some events detected in the region R1 are not the microbial cells but the particles produced during the disintegration of the granules. FSC for the stained cells was higher for the non-stained cells.

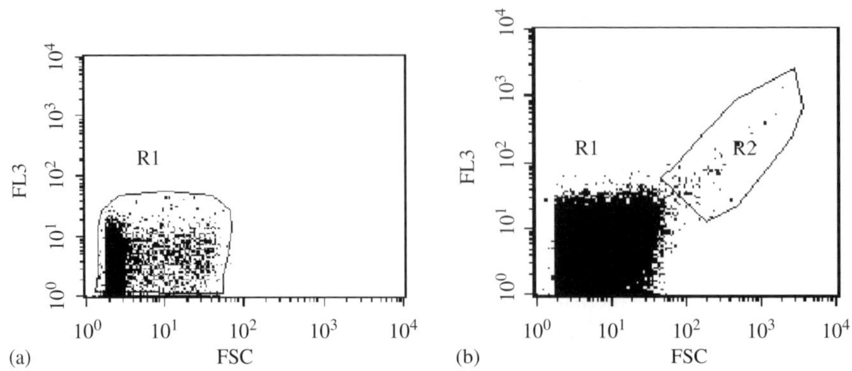

Fig. 7.6. Dot plot of red fluorescence (FL3) and forward light scatter (FSC) of the cells, which were not incubated with the probe (a, control) and incubated with Bacto1080 probe (b, experiment). The cells were produced by the disintegration of big granules. The region R1 is corresponding to non-labeled cells and particles and R2 region shows the cells, which were hybridized with Bacto1080 probe.

Flow cytometry, additionally to SCLM data, proves the presence of *Bacteroides spp.* in the granules and the bigger share of obligate anaerobic bacteria in big granules rather than in small granules.

All methods confirmed that aerobically grown microbial granules contain obligate anaerobe *Bacteroides spp.* It is well known that these bacteria usually dominate in human feces. Probably, the source of the initial inoculation of aerobic microbial flocs in wastewater treatment plant by *Bacteroides spp.* was the raw sewage or the effluent from anaerobic digester. Ecological consequence from the detection of obligate anaerobic bacteria in aerobically grown granule is that the boundary of anaerobic microzone in microbial matrix of the granule or thick biofilm can be detected by the visualization of the layer of *Bacteroides spp.* or other anaerobic microorganisms by FISH with specific fluorescence-labeled oligonucleotide probe. This method reliably shows the place of anaerobic layer, which is depending on average oxygen gradient during the microbial granule growth and activity. If the place of anaerobic zone is detected by the microelectrodes, it could be depended on the variable oxygen gradient in the granule during the measurement. Applied significance of the detection of obligate anaerobes in aerobically grown granule is that there must be optimal size of such granules to ensure optimal balance of aerobic and anaerobic biodegradation of organic matter in the granule.

Fig. 7.7. Floating of the granules after settling. (a) to (e) correspond to 2, 24, 25, 28, and 32 min after settling of the granules.

The produced gases of fermentation can destroy the granule if the size of the granule is big and, correspondingly, the number of anaerobic bacteria inside the granule will be high.

Another negative effect of anaerobic bacteria on the wastewater treatment process is the occurrence of floating granules, which could occur if anaerobic bacteria, and especially denitrifying bacteria, are allowed to incubate in medium with carbon source after biomass settling (Fig. 7.7). This potential floating of the microbial granules in case of high organic or nitrate load, leading to the production of gases in anaerobic zone of the granule, can deteriorate wastewater treatment.

Microbial Diversity of Granules, Grown in Glucose-containing Model Wastewater, Studied by FISH with Group-specific Oligonucleotide Probes

Bacterial populations, associated with the development of aerobic granules in glucose-containing model wastewater, were monitored by a combination of FISH with rRNA-targeted fluorescence *in situ* hybridization with group specific probes and CLSM (Table 7.5). A column-type sequential aerobic sludge blanket reactor (SASBR) was inoculated with activated sludge from a municipal plant and fed with synthetic wastewater containing glucose as the main carbon source. The reactor biomass initially consisted of flocs that eventually developed into well-settling granules.

One of the early models of aerobic granulation involved the growth and subsequent fragmentation of filamentous fungal pellets under stressed conditions (Beun et al., 1999, 2000). These fungal fragments functioned as an immobilization matrix for colonization by floc-forming bacteria, and eventually grew out to become granules with good settling properties that were easily retained in the reactor. Based on microscopy images, Beun et al. (1999) observed the proliferation of filamentous fungal pellets during the initial stage of operation of a sequential batch reactor. These mycelial pellets would eventually expand and lyze as a result of oxygen limitation. The resulting fragments would then serve as an immobilization matrix for bacteria to attach and form microcolonies that would in turn grow into compact granules. The filamentous bacteria in the current study played a similarly important role to the filamentous fungi in the formation of aerobic granules, by facilitating floc aggregation and providing attachment sites for colonization by other microorganisms. To obtain counts of filamentous bacteria in our experiments, 50 μl of mixed liquor sample was transferred to a clean glass slide and covered with a cover slip. An eyepiece with a single hairline was used to scan the length of the cover slip at 100× magnification. Eleven fields of view were chosen, and for each field of view, the number of times that filamentous bacteria intersected the hairline was counted (Jenkins et al., 1993; Seviour et al., 1999). Filamentous bacteria in sludge samples were identified and categorized into taxonomic groups according to the schemes of Eikelboom and van Buisen (1981) and Jenkins et al. (1993) based on microscopy observations. Gram staining was performed using the modified Hucker method (Seviour et al., 1999).

After reactor start-up, the filament counts increased significantly from $329 \pm 26\ \mu l^{-1}$ on day 2 to $406 \pm 33\ \mu l^{-1}$ on day 6 and then decreased to $91\ \mu l^{-1}$ on day 16. Filamentous bacteria that fit the morphological descriptions of *Sphaerotilus natans* and Type 1701 were observed in both flocs and granules. Types 1851 and 1863 were also observed, but to a lesser extent. Filamentous bacteria probably played a role in granule formation by acting as bridges to interconnect the sludge flocs and serving as a matrix

Table 7.5. Group specific oligonucleotide probe sequences and target sites

Probe	Sequence (5′–3′)	rRNA target site	Specificity	Percentage of formamide used	Reference
EUB 338	GCTGCCTCCCGTAGGAGT	16S, 338–355	Bacteria	20	Amann et al., 1995, 1998
EUB 338-II	GCAGCCACCCGTAGGTGT	16S, 338–355	Bacteria	20	Daims et al., 1999
EUB 338-III	GCTGCCACCCGTAGGTGT	16S, 338–355	Bacteria	20	Daims et al., 1999
ALF1b	CGTTCG(C/T)TCTGAGCCAG	16S, 19–35	α-*Proteobacteria*	20	Manz et al., 1992
BET42a	GCCTTCCCACTTCGTTT	23S, 1027–1043	β-*Proteobacteria*	35	Manz et al., 1992
GAM42a	GCCTTCCCACATCGTTT	23S, 1027–1043	γ-*Proteobacteria*	35	Manz et al., 1992
HGC69a	TATAGTTACCACCGCCGT	23S, 1901–1918	*Actinobacteria* with high DNA G+C content	25	Roller et al., 1994
ARCH915	GTGCTCCCCCGCCAATTCCT	16S, 915–934	Archaea	20	Stahl and Amann, 1991
CF319a	TGGTCCGTGTCTCAGTAC	16S, 319–336	*Cytophaga-Flavobacterium*	35	Manz et al., 1996
LGC354A	TGGAAGATTCCCTACTGC	16S, 354–371	*Actinobacteria* with low DNA G+C content	35	Meier et al., 1999
LGC354B	CGGAAGATTCCCTACTGC	16S, 354–371	*Actinobacteria* with low DNA G+C content	35	Meier et al., 1999
LGC354C	CCGAAGATTCCCTACTGC	16S, 354–371	*Actinobacteria* with low DNA G+C content	35	Meier et al., 1999

for attachment of floc-forming bacteria and subsequent granule formation (Figs 7.8–7.10).

Gram-positive bacteria with high G+C content, β-*Proteobacteria*, and *Cytophaga-Flavobacteria* were the dominant bacterial sub-populations in flocs sampled on day 2, and their relative abundance increased significantly in granules sampled on day 23 (Figs 7.11 and 7.12). Gram-positive bacteria with low G+C content and γ- and α-*Proteobacteria* were minor constituents in the flocs and granules, while Archaea were not detected at all. The development of granules from flocs was characterized by a remarkable increase in both the cell count and area of cell coverage of high G+C gram-positive bacteria, β-*Proteobacteria*, and *Cytophaga-Flavobacteria*. These three groups of bacteria also represented the dominant sub-populations in the flocs and granules. On the other hand, members of γ-*Proteobacteria*, α-*Proteobacteria,* and low G+C gram-positive bacteria represented minor sub-population in the flocs and granules. In addition, no significant changes in relative abundances were detected for these three bacterial groups as the flocs developed into granules.

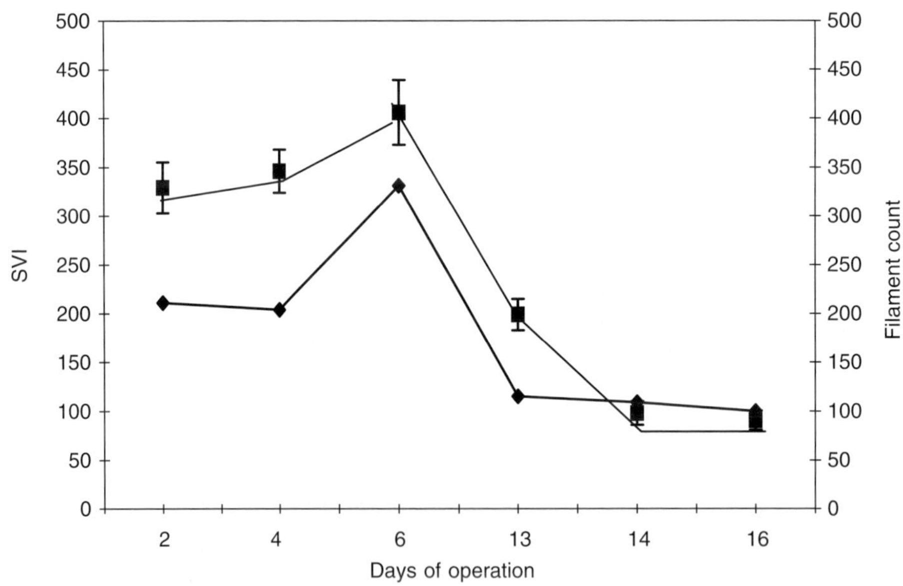

Fig. 7.8. Profile of filament counts (■, filaments μl^{-1}) and SVI (♦, ml g^{-1}) during reactor operation.

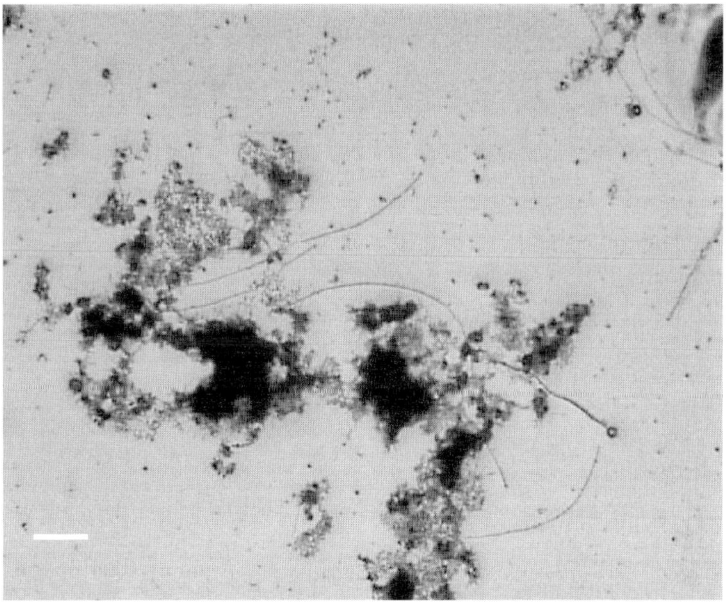

Fig. 7.9. Filament bridging of aerobic flocs. Scale bar represents 10 μm.

Gram-positive bacteria with low G+C content were detected at relatively low abundances in this study with an equimolar mixture of LGC354A, B, and C as recommended by Meier et al. (1999). In contrast, gram-positive bacteria with high G+C content constituted in excess of 24% of Eubacteria by both cell count and cell area measures in the day 2 flocs, and this percentage exceeded 33% in day 23. Gram-positive bacteria with low G+C content includes several genera that are present in the gastrointestinal tracts and feces of humans (McCartney et al., 1996), and this group of bacterial typically constitutes a minor fraction of the bacterial population in normal conventional activated sludge systems. However, they are known to proliferate in environments with high concentrations of inorganic and organic nitrogen, such as treatment plants that process livestock wastewater or aquifers that are polluted by livestock wastewater (Cho and Kim, 2000). Several gram-positive bacteria with high G+C content isolated from activated sludge plants are known to consume soluble COD rapidly and store them as storage polymers such as glycogen (Nakamura et al., 1995; Maszenan et al., 1999, 2000a; Liu et al., 2001). This competitive advantage allows them to thrive in environments with

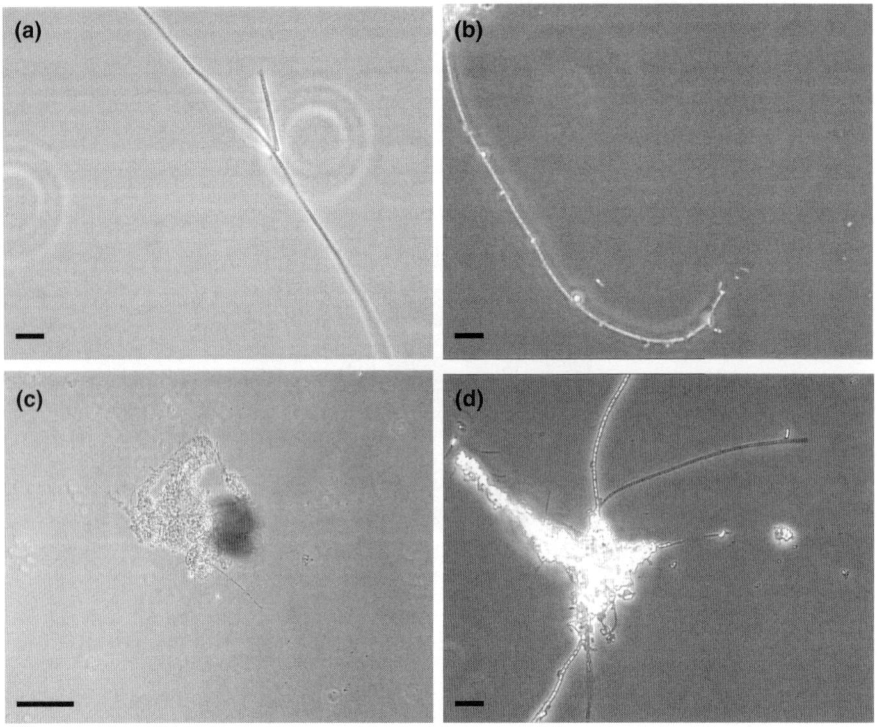

Fig. 7.10. Light microscopy images of aerobic flocs containing Sphaerotilus natans (a), and Types 1701 (b), 1851 (c), and 1863 (d). Scale bars for a, b, and d represent 10 μm. Scale bar for c represents 100 μm.

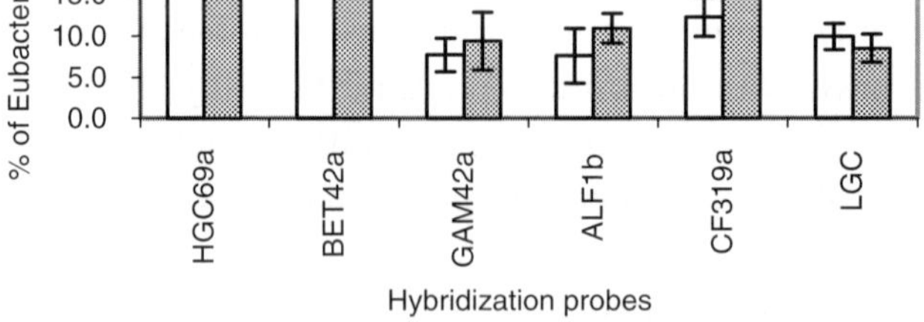

Fig. 7.11. Bacterial population enumeration using fluorescent *in situ* hybridization probes on day 2 (□) and day 23 (▩) based on cell counts.

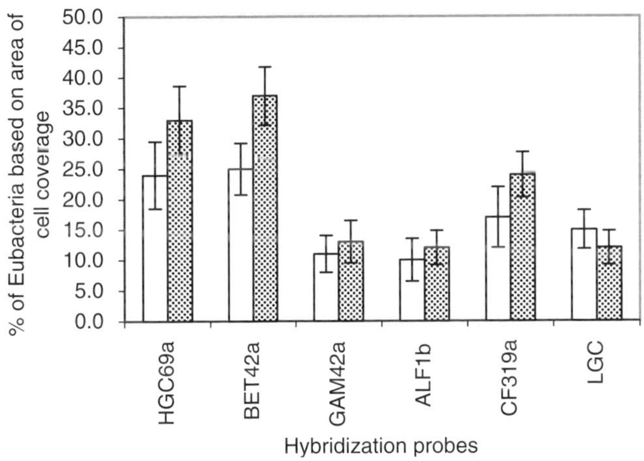

Fig. 7.12. Bacterial population enumeration using fluorescent *in situ* hybridization probes on day 2 (□) and day 23 (▩) based on area of cell coverage.

a low food-to-microorganism ratio. These environments would include activated sludge systems, such as the one from which seed sludge was obtained for this study, and sequential batch reactors, such as the one used in this study.

The dominance of β-*Proteobacteria* in the aerobic granulation process is not surprising as this group of microorganisms, which includes the genera *Comamonas, Hydrogenophaga,* and *Acidovorax*, are frequently found in both natural and engineered systems as they are nutritionally versatile and can consume a wide array of carbon substrates (Snaidr et al., 1997). β-*Proteobacteria* have also been shown to dominate the attached bacterial community during the initial development of river biofilms in a rotating annular reactor system (Manz et al., 1999). Based on this study, aerobic granulation which is another form of biofilm, is the growth of self-immobilizing bacteria occurring without supporting matrix. Filamentous bacteria fitting the morphotype description of *S. natan* which also hybridized with BET42a probe, with bacteria attached may have an important role in granulation. This observation is also supported by the isolation of bacterial strains growing on the sheath of *S. natans*, and producing enzymes capable of degrading the sheath polysaccharides moiety (Takeda et al., 2002).

Bacterial Populations in Acetate-fed Aerobic Granules

Bacterial populations associated with the development of aerobic granules in column-type sequential aerobic sludge blanket reactors fed with a synthetic acetate-based wastewater were monitored also by microscopy and ribosomal RNA-targeted fluorescence *in situ* hybridization (FISH) using group specific probes. The reactor biomass initially consisted of pin-point flocs which later developed into well-settling granules. Filament counts and SVI decreased rapidly from $389 \pm 26\ \mu l^{-1}$ and $341\ ml\ g^{-1}$ on day 2 to $14 \pm 26\ \mu l^{-1}$ and $111\ ml\ g^{-1}$on day 16. Filamentous bacteria fitting the morphological descriptions of *Sphaerotilus natans* and Type 1863 were observed on day 2 in sludge flocs after seeding of reactor. However, filament counts decreased due to rapid filament washing out and proliferation of floc-forming bacteria.

FISH determination revealed that gram-positive bacteria with high G+C content and β-*Proteobacteria* were the dominant sub-populations in flocs, and their relative abundances increased in granules sampled on day 23. Gram-positive bacteria with low G+C content, α-, and γ-*Proteobacteria* were minor constituents in both floc and granules, while Archaea were not detected at all.

References

Amann, R., & Kühl, M. (1998). *In situ* methods for assessment of microorganisms and their activities. *Curr. Opin. Microbiol.*, *1*, 352–358.

Amann, R., Ludwig, W., & Schleifer, K.H. (1995). Phylogenetic identification and in situ detection of individual microbial cells without cultivation. *Microbiol. Rev.*, *59*, 143–169.

Amann, R., Lemmer, H., & Wagner, M. (1998). Monitoring the community structure of wastewater treatment plants: a comparison of old and new techniques. *FEMS Microbiol. Ecol.*, *25*, 205–215.

Beun, J.J., Hendriks, A., van Loosdrecht, M.C.M., Morgenroth, E., Wilderer, P.A., & Heijnen, J.J. (1999). Aerobic granulation in a sequencing batch reactor. *Water Res.*, *33*, 2283–2290.

Beun, J.J., van Loosdrecht, M.C.M., & Heijnen, J.J. (2000). Aerobic granulation. *Water Sci. Technol.*, *41*, 41–48.

Chandler, D.P., Brockman, F.J., & Fredrickson, J.K. (1997). Use of 16S rDNA clone libraries to study changes in a microbial community resulting from *ex situ* perturbation of a subsurface sediment. *FEMS Microbiol. Rev.*, *20*, 217–230.

Cho, J.C., & Kim, S.J. (2000). Increase in bacterial community diversity in subsurface aquifers receiving livestock wastewater input. *Appl. Environ. Microbiol.*, *66*, 956–965.

Daims, H., Bruhl, A., Amann, R., Schleifer, K.-H., & Wagner, M. (1999). The domain-specific probe EUB338 is insufficient for the detection of all bacteria: development and evaluation of a more comprehensive probe set. *Syst. Appl. Microbiol.*, *22*, 434–444.

Dang, H., & Lovell, C.R. (2000). Bacterial primary colonization and early succession on surfaces in marine water as determined by amplified rRNA gene restriction analysis and sequence analysis of 16S rRNA genes. *Appl. Environ. Microbiol.*, *66*, 467–475.

Dojka, M.A., Hugenholtz, P., Haack, S.K., & Pace, N.R. (1998). Microbial diversity in a hydrocarbon- and chlorinated-solvent-contaminated aquifer undergoing intrinsic bioremediation. *Appl. Environ. Microbiol.*, *64*, 3869–3877.

Dore, J., Sghir, A., Hannequart-Gramet, A., Gorthier, G., & Pochart, P. (1998). Design and evaluation of a 16S rRNA-targeted oligonucleotide probe for specific detection and quantification of human faecal *Bacteroides* populations. *Syst. Appl. Microbiol.*, *1*, 65–71.

Eikelboom, D.H., & van Buisen, H.J. (1981). *Microscope Sludge Investigation Manual.* TNO Research Institute for Environmental Hygiene, Report A94a Delft, Netherlands.

Farrelly, V., Rainey, F.A., & Stackebrandt, E. (1995). Effect of genome size and rrn gene copy number on PCR amplification of 16S rRNA genes from a mixture of bacterial species. *Appl. Environ. Microbiol.*, *61*, 2798–2801.

Gich, F.B., Amer, E., Figueras, J.B., Abella, C.A., Balaguer, M.D., & Poch, M. (2000). Assessment of microbial community structure changes by amplified ribosomal DNA restriction analysis (ARDRA). *Int. Microbiol.*, *3*, 103–106.

Head, I.M., Saunders, J.R., & Pickup, R.W. (1998). Microbial evolution, diversity, and ecology: a decade of ribosomal RNA analysis of uncultivated microorganisms. *Microb. Ecol.*, *35*, 1–21.

Jenkins, D., Richard, M.G., & Daigger, G.T. (1993). *Manual on the Causes and Control of Activated Sludge Bulking and Foaming*. Lewis Publishers, Chelsea, Michigan.

Liu, W.-T., Nielsen, A.T., Wu, J.-H., Tsai C.-S., Matsuo, Y., & Molin, S. (2001). In situ identification of polyphosphate- and polyhydroxyalkanoate-accumulating traits for microbial populations in a biological phosphorus removal process. *Environ. Microbiol.*, *3*, 110–122.

Manz, W., Amann, R., Ludwig, W., Wagner, M., & Schleifer, K.-H. (1992). Phylogenetic oligodeoxynucleotide probes for the major subclasses of *Proteobacteria*: Problems and solutions. *Syst. Appl. Microbiol., 15*, 593–600.

Manz, W., Szewzyk, U., Ericsson, P., Amann, R., Schleifer, K.H., & Stenstrom, T.A. (1993). *In situ* identification of bacteria in drinking water and adjoining biofilms by hybridization with 16S and 23S rRNA-directed fluorescent oligonucleotide probes. *Appl. Environ. Microbiol., 59*, 2293–2298.

Manz, W., Amann, R., Ludwig, W., Vancanneyt, M., & Schleifer, K.-H. (1996). Application of a suite of 16S rRNA-specific oligonucleotide probes designed to investigate bacteria of the *phylum cytophaga-flavobacter-bacteriodes* in the natural environment. *Microbiol., 142*, 1097–1106.

Manz, W., Wendt-Potthoff, K., Neu, T.R., Szewzyk, U., & Lawrence, J.R. (1999). Phylogenetic composition, spatial structure, and dynamics of lotic bacterial biofilms investigated by fluorescent *in situ* hybridization and confocal laser scanning microscopy. *Microb. Ecol., 37*, 225–237.

Maszenan, A.M., Seviour, R.J., Patel, B.K.C., Schumann, P., & Rees, G.N. (1999). *Tessaracoccus bendigoensis* gen. nov, sp. nov., a gram-positive coccus occurring in regular packages or tetrads, isolated from activated sludge biomass. *Int. J. Syst. Bacteriol., 49*, 459–468.

McCartney, A.L., WenZhi, W., & Tannock, G.W. (1996). Molecular analysis of the composition of the *Bifidobacterial* and *Lactobacillus* microflora of Humans. *Appl. Environ. Microbiol., 62*, 4608–4613.

Meier, H., Amann, R., Ludwig, W., & Schleifer, K.-H. (1999). Specific oligonucleotide probes for in situ detection of a major group of gram positive bacteria with low DNA G+C content. *Syst. Appl. Microbiol., 22*, 186–196.

Moyer, C.L., Dobbs, F.C., & Karl, D.M. (1994). Estimation of diversity and community structure through restriction fragment length polymorphism distribution analysis of bacterial 16S rRNA genes from a microbial mat at an active, hydrothermal vent system, Loihi Seamount, Hawaii. *Appl. Environ. Microbiol., 60*, 871–879.

Moyer, C.L., Tiedje, J.M., Dobbs, F.C., & Karl, D.M. (1996). A computer-simulated restriction fragment length polymorphism analysis of bacterial small-subunit rRNA genes: efficacy of selected tetrameric restriction enzymes for studies of microbial diversity in nature. *Appl. Environ. Microbiol., 62*, 2501–2507.

Nakamura, K., Hiraishi, A., Yoshimi, Y., Kawaharasaki, M., Masuda, K., & Kamagata, Y. (1995). *Microlunatus phosphovorus* gen. nov., sp. nov., a new gram-positive polyphosphate-accumulating bacterium isolated from activated sludge. *Int. J. Syst. Bacteriol., 45*, 17–22.

Olsen, G.J., Matsuda, H., Hagstrom, R., & Overbeek, R. (1994). FastDNAml: a tool for construction of phylogenetic trees of DNA sequences using maximum likelihood. *Comput. Appl. Biosci.*, *10*, 41–48.

Roller, C., Wagner, M., Amann, R., Ludwig, W., & Schleifer, K.-H. (1994). *In situ* probing of Gram-positive bacteria with high DNA G+C content using 23S rRNA-targeted oligonucleotides. *Microbiology*, *140*, 2849–2858.

Seviour, E.M., Seviour, R.J., & Lindrea, K.C. (1999). Descriptions of the filamentous bacteria causing bulking and foaming in activated sludge. *The Microbiology of Activated Sludge* (eds. Seviour, R.J., & Blackall, L.L.), Kluwer Academic Publishers, Dordrecht, The Netherlands, 301–347.

Sghir, A., Gramet, G., Suau, A., Rochet, V., Pochart, P., & Dore, J. (2000). Quantification of bacterial groups within human fecal flora by oligonucleotide probe hybridization. *Appl. Environ. Microbiol.*, *66*, 2263–2266.

Snaidr, J., Amann, R., Huber, I., Ludwig, W., & Schleifer, K.-H. (1997). Phylogenetic analysis and in situ identification of bacteria in activated sludge. *Appl. Environ. Microbiol.*, *63*, 2884–2896.

Stahl, D., & Amann, R. (1991). Development and application of nucleic acid probes in bacterial systematics. *Nucleic Acid Techniques in Bacterial Systematics* (eds. Stackebrandt, E., & Goodfellow, M.), John Wiley & Sons Ltd., Chichester, England, 205–248.

Suzuki, M.T., & Giovannoni, S.J. (1996). Bias caused by template annealing in the amplification of mixtures of 16S rRNA genes by PCR. *Appl. Environ. Microbiol.*, *62*, 625–630.

Takeda, M., Kamagata, Y., Shinmaru, S., Nishiyama, T., & Koizumi, J.I. (2002). *Paenibacillus koleovorans* sp. nov., able to grow on the sheath of *Sphaerotilus natans*. *Int. J. Syst. Bacteriol.*, *57*, 1597–1601.

Urakawa, H., Kita-Tsukamoto, K., & Ohwada, K. (1999). Microbial diversity in marine sediments from Sagmi bay and Tokyo bay, Japan, as determined by 16S rRNA gene analysis. *Microbiology*, *145*, 3305–3315.

von Wintzingerode, F., Gobel, U.B., & Stackebrandt, E. (1997). Determination of microbial diversity in environmental samples: pitfalls of PCR-based rRNA analysis. *FEMS Microbiol. Rev.*, *21*, 213–229.

Chapter 8

Nutrient Removal by Microbial Granules

Yu Liu

Introduction

Nitrogen and phosphorus are key nutrients causing eutrophication in water body, and they are required to be removed from water resources in many countries. Biological nutrient removal (BNR) can be accomplished by nitrification, denitrification, and enhanced biological phosphorus removal (EBPR). However, these systems often encounter several difficulties including the sludge bulking due to proliferation of filamentous organisms, requirement for a long sludge age to ensure stable nitrification as well as requirement for a large space if multistage systems are employed.

In the past few years, aerobic granules for organic carbon removal had been developed in sequencing batch reactors (SBR) (Beun et al., 1999; Peng et al., 1999; Etterer and Wilderer, 2001; Tay et al., 2001). Compared to conventional activated sludge flocs, aerobic granular sludge has regular, dense, and strong physical structure, good settling ability, high biomass retention, and the ability to withstand shock load. Recently, research attention has been given to the development of granules capable of removing nitrogen and phosphorus (Beun et al., 2001; Lin et al., 2003; Yang et al., 2003; de Kreuk and van Loosdrecht, 2004; Qin et al., 2005). Evidence shows that the problems encountered in the suspended growth nutrient-removal system, such as sludge bulking, large treatment plant space, washout of nitrifying biomass, secondary P release in a clarifier, higher production of waste sludge, would be overcome by developing and applying N-removing and P-accumulating granules (Lin et al., 2003; Yang et al., 2003). A more compact and efficient granule-based biotechnology would be expected for high-efficiency N and P removal. Therefore, this chapter presents up-to-date information with regard to the development of microbial granules for the nutrient removal processes.

Development of Nitrifying Granules

Yang et al. (2005) cultivated nitrifying granules in at various substrate N/COD ratios in the range of 5/100–30/100 by weight. It was found that nitrifying granules could form at all the tested substrate N/COD ratios and the characteristics of nitrifying granules were found to be substrate N/COD ratio-dependent.

The Formation of Nitrifying Granules

Nitrifying granulation was found to be a gradual process from the dispersed seed sludge through the tiny aggregates to the mature spherical granules as shown in Fig. 8.1 (Yang et al., 2005). After about four weeks of operation, nitrifying granules with a dense structure formed at the substrate N/COD ratios studied. The biomass concentrations in the reactors were increased up to 10 g/l of total solids (TS) upon the formation of nitrifying granules. The granules developed at all tested N/COD ratios exhibit compact structure compared to the seed sludge (Fig. 8.1).

Characteristics of Nitrifying Granules

Granule Size

Compared to the seed activated sludge with a floc size of 0.09 mm, mature nitrifying granules had a mean size of 1.9, 1.5, 0.5, and 0.4 mm in the reactors operated at substrate N/COD of 5/100–30/100, respectively (Fig. 8.1). These seem to indicate that small nitrifying granules would be cultivated at high substrate N/COD ratios and vice versa.

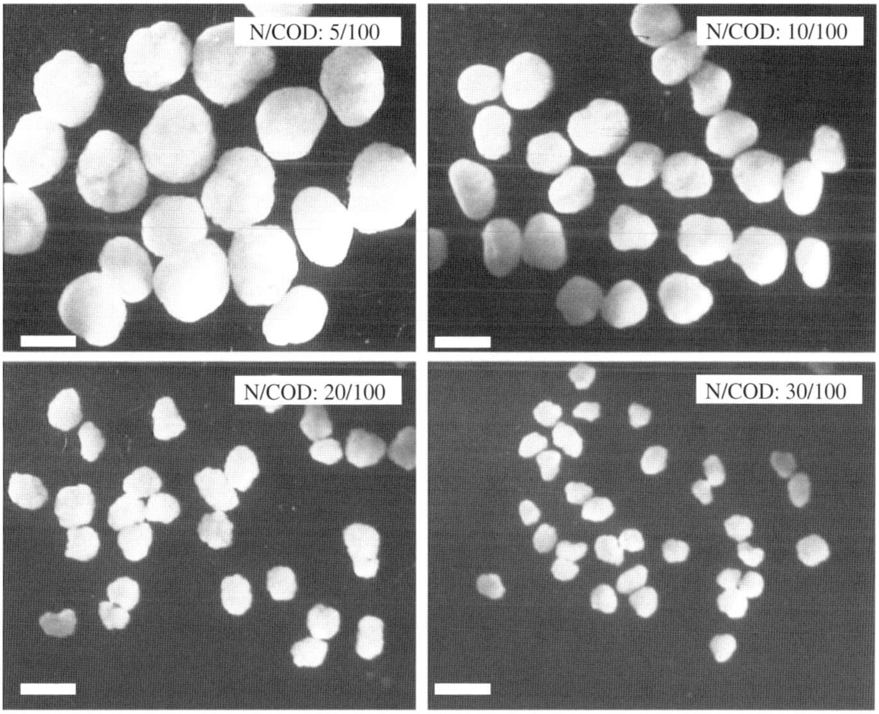

Fig. 8.1. Morphology of mature nitrifying granules developed at different substrate N/COD ratios. Bar: 1 mm (Yang et al., 2005).

Settleability

In the environmental engineering field, sludge settleability has been commonly described by sludge volume index (SVI). Figure 8.2 shows changes of SVI in the course of granulation observed at the different substrate N/COD ratios (Yang et al., 2005). It appears that the SVI of microbial granules exhibited a decreasing trend with the increase of substrate N/COD ratio, e.g. the lowest SVI was found at the highest substrate N/COD ratio of 30/100. These results imply that the substrate N/COD ratio would have a significant effect on the structure of microbial granules, i.e. a more compact microbial structure could be expected at higher substrate N/COD ratio. Such information is important because one may expect to manipulate the structure of aerobic granules by adjusting substrate N/COD ratios.

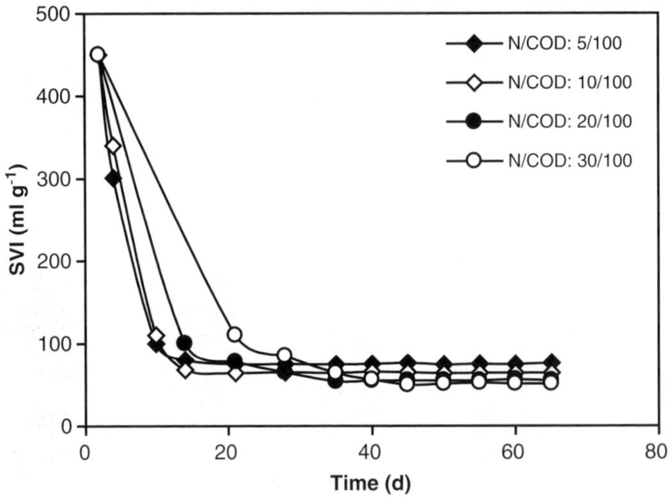

Fig. 8.2. Change of SVI during aerobic granulation at different substrate N/COD ratios (Yang et al., 2005).

In addition, compared to the seed activated sludge with a SVI of 265 ml/g, the settleability of microbial granules was improved noticeably.

It should be realized that the settling velocity of aerobic granules cultivated at different substrate N/COD ratios was greater than 60 m/h (Yang et al., 2005), while the settling velocity of conventional activated sludge was around 5 m/h (Giokas et al., 2003). Compared to the conventional activated sludge flocs, the excellent settleability of aerobic granules can ensure quick and effective separation of biosolids from the effluent and high biomass retention can be achieved in the reactors. These are very attractive from the point of view of industrial application as they may help to solve the problems encountered in conventional nitrogen removal processes, such as sludge bulking, washout of nitrifying biomass, and so on.

Specific Gravity

The specific gravity can be used to describe the compactness of a microbial community. Figure 8.3 shows the specific gravities of the mature aerobic granules developed at different substrate N/COD ratios (Yang et al., 2003). It can be seen that the specific gravity of granules tends to increase with

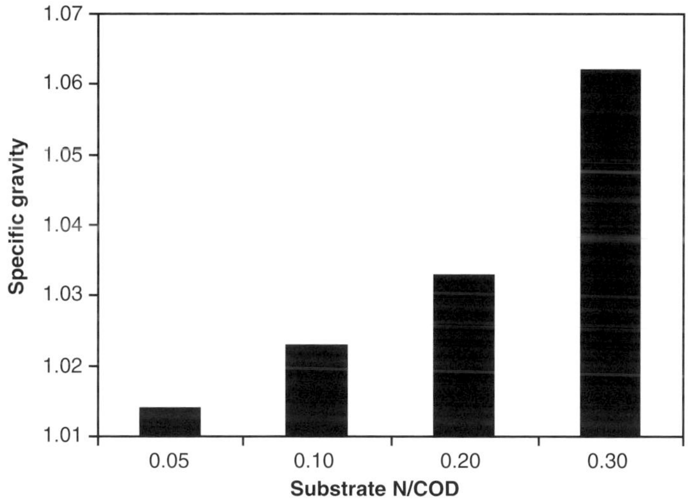

Fig. 8.3. Specific gravity of aerobic granules developed at different substrate N/COD ratios (Yang et al., 2003).

the increase of substrate N/COD ratio, i.e. a high substrate N/COD ratio would result in a more compact structure of aerobic granules. Such a trend is indeed consistent with the changes in SVI as illustrated in Fig. 8.2. Compared to the seed sludge with a specific gravity of 1.002, it is obvious that the aerobic granules had a much denser and more compact microbial structure.

Cell Hydrophobicity

Changes in cell hydrophobicity of microbial granules cultivated at different substrate N/COD ratios are presented in Fig. 8.4 (Yang et al., 2005). The cell hydrophobicity gradually increased until a stable value was achieved after a 40-day operation, while the cell hydrophobicity at steady state exhibits a positive relation to the substrate N/COD ratio. Increasing evidence shows that cell hydrophobicity plays a crucial role in the formation of biofilm, anaerobic granules as well as aerobic granules (Mahoney et al., 1987; Rouxhet and Mozes, 1990; Tay et al., 2000; Liu et al., 2004a). In a thermodynamic sense, the increase of cell hydrophobicity would simultaneously cause a decrease in the excess Gibbs energy of the surface, which in turn favors the self-aggregation of bacteria from liquid

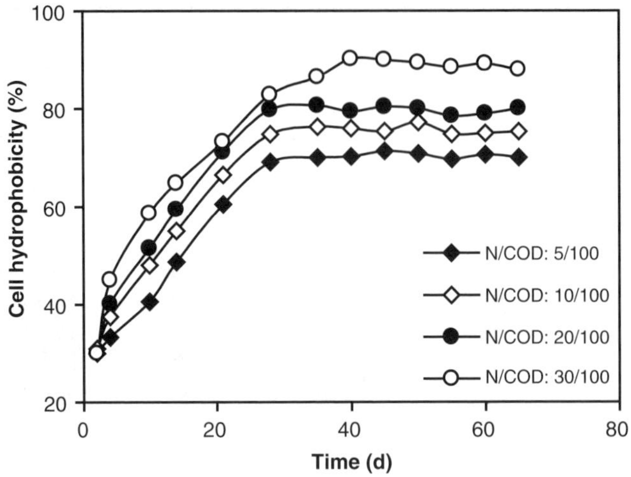

Fig. 8.4. Changes in cell hydrophobicity during aerobic granulation at different substrate N/COD ratios (Yang et al., 2005).

phase to form a new solid phase, namely microbial aggregates (Liu and Tay, 2002). In fact, the cell hydrophobicity of bacteria has been recognized as a decisive affinity force in cell immobilization process (Rouxhet and Mozes, 1990; Bossier and Verstraete, 1996; Zita and Hermansson, 1997; Liu et al., 2004a). It seems certain that hydrophobic binding force has a prime importance for the cell-to-cell approach and interaction, and the hydrophobicity of bacterial surface can act as a driving force for the initiation of cell-to-cell aggregation, which is the first step towards aerobic granulation, and further keep bacteria aggregated tightly together.

It should be pointed out that increased substrate N/COD ratio would lead to an enriched nitrifying population in aerobic granules. It had been shown that nitrifying bacteria had a higher hydrophobic interaction than that of activated sludge microorganisms (Sousa et al., 1997; Kim et al., 2000). Thus, high cell hydrophobicity of aerobic granules developed at high substrate N/COD ratio, in part, could be attributed to the enriched nitrifying population in granules.

Production of Extracellular Polysaccharides

Extracellular polysaccharides can mediate both cohesion and adhesion of cells, and play a crucial role in building and maintaining structural

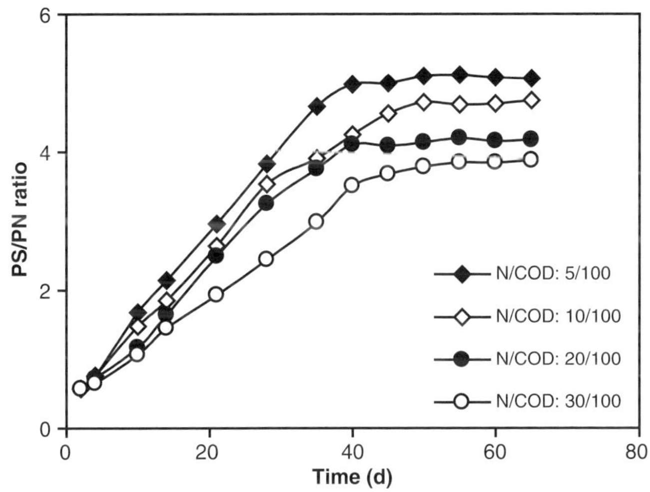

Fig. 8.5. Changes in the PS/PN ratio in the course of aerobic granulation at different substrate N/COD ratios (Yang et al., 2005).

integrity in a community of immobilized cells (Fletcher and Floodgate, 1973; Schmidt and Ahring, 1996; Lopes et al., 2000; Liu et al., 2004b; Wang et al., 2005). Figure 8.5 shows changes in the ratio of extracellular polysaccharides (PS) to extracellular proteins (PN) in the course of aerobic granulation at different substrate N/COD ratios (Yang et al., 2005). The salient points from Fig. 8.5 can be summarized as follows: (i) the PS/PN ratio increased in a very significant way with the formation of aerobic granules, e.g. the PS/PN ratio increased from an initial value of 0.57 for the seed sludge to 4.0–5.0 for the aerobic granules. These may indicate that microbial aggregation would be partially related to the production of extracellular polysaccharides; (ii) with increasing the substrate N/COD ratio, the PS/PN ratio shows a decreasing trend. In fact, this is in line with previous finding showing that a reduced substrate N/COD ratio would stimulate the production of extracellular polysaccharides, resulting in improved bacterial attachment to solid surfaces (Schmidt and Ahring, 1996; Durmaz and Sanin, 2001). In addition, Tsuneda et al. (2003) found that extracellular polysaccharides exhibited good correlation with cell adhesion, while no protein was related to cell adhesion.

Extracellular polysaccharides are produced by most bacteria out of cell wall with the purpose of providing cells with the ability to compete in a variety of environments, providing a mode for adhesion to surfaces

(Sutherlan, 2001). The PS/PN ratios of aerobic granules tend to decrease with the increase of the substrate N/COD ratio (Fig. 8.5). As the nitrifying populations in aerobic granules were greatly enriched at high substrate N/COD ratios, the lower production of extracellular polysaccharides at higher substrate N/COD ratio can be explained in a way such that nitrifying bacteria cannot utilize organic carbon for microbial growth, and only 11–27% of energy generated goes to biosynthesis (Laudelout et al., 1968). Thus, less extracellular polysachharides would be synthesized in aerobic granules developed at high substrate N/COD ratio.

It was also found that the production of cell polysaccharides is quasi-linearly dependent on the respiration activity of heterotrophic bacteria present in the aerobic granules, i.e. a high catabolic activity favors the production of cell polysaccharides. These are consistent with other research showing that the production of extracellular polysaccharides was energy-dependent (Robinson et al., 1984; Wuertz et al., 1998). In fact, there is evidence that cell carbohydrate content increased and protein content decreased significantly, the way as the substrate N/COD ratio decreased (Durmaz and Sanin, 2001). It seems reasonable to consider that nitrifying bacteria would produce much less extracellular polysaccharides than heterotrophs. In addition, Tsuneda et al. (2001) used extracellular polysaccharides produced by heterotrophic bacteria to enhance the formation of nitrifying biofilm. As shown in Fig. 8.5, the content of aerobic granule-polysaccharides at steady state was at about 3-fold higher than that of proteins. Vandevivere and Kirchman (1993) also found that the content of exopolysaccharides was 5-fold greater for attached cells than for free-living cells. These might imply that cell proteins would less contribute to the structure and stability of immobilized microorganisms. A more recent research further showed that the structural stability of aerobic granules was closely related to the content and distribution of insoluble extracellular polysaccharides (Wang et al., 2005).

Elemental Compositions of Nitrifying Granules

The characteristics of aerobic granules seem to be substrate N/COD ratio-dependent. It has been known that changes in characteristics are usually related to the changes in chemical compositions of microorganisms (Pitryuk et al., 2002). In mixed microbial culture, chemical compositions of microorganisms may reflect the changes of microbial community and growth conditions (Duboc et al., 1995; Heldal et al., 1996). Table 8.1 shows the elemental compositions of aerobic granules developed at different substrate N/COD ratios (Liu et al., 2003a). These data indicate that aerobic granules mainly comprised six major elements, i.e. C, H, O, N, S, and P. The substrate N/COD ratio displays a profound effect on the respective ratio of cell oxygen, nitrogen, and calcium normalized to cell carbon. Cell N/C ratio increased with the increase in the substrate N/COD ratio, whereas cell O/C ratio decreased. Heldal et al. (1996) observed a marked reduction in cell O/C level when the conditions changed from nitrogen-limitation to carbon-limitation. Vrede et al. (2002) also reported that elemental composition of bacterioplankton was closely related to the substrate carbon and nitrogen and the lowest cell carbon content was found in carbon-limited cells. Microorganisms are often found to differ in their relative contents of C, H, N, O, and other elements when they experience the shift of microbial community and the change of growth conditions (Duboc et al., 1995; Pitryuk et al., 2002). Obviously, information on chemical composition of microorganisms is essential for a sound understanding of the behaviors of microbial community.

The accumulation of calcium in anaerobic granules had been reported (Yu et al., 2001). However, it appears from Table 8.1 that no accumulation of cell calcium occurred in aerobic granules cultivated at different substrate N/COD ratios. In fact, the cell Ca/C ratio of aerobic granules is even lower than that of the seed sludge (7.5 mmol mol^{-1}). Increasing evidence shows that the accumulation of calcium in aerobic granules would be related to organic substrate used, e.g. aerobic granules reported Table 8.1 were grown on ethanol, however, aerobic granule grown on acetate had high calcium content (Qin et al., 2004a).

Microbial Diversity of Nitrifying Granules

To remove organics and nitrogen from wastewater, nitrifying, denitrifying, and heterotrophic populations should co-exist in microbial granules. It had been reported that substrate with different N/COD ratios would lead to significant shift among various populations in both suspended and attached cultures (Moreau et al., 1994; Ohashi et al., 1995; Prinčič et al., 1998; Ballinger et al., 2002). A variation of the relative substrate composition

Table 8.1. Elemental compositions of aerobic granules developed at different substrate N/COD ratios, in terms of percentage by dry weight of granules (Liu et al., 2003a)

Element	Substrate N/COD ratio			
	5/100	10/100	20/100	30/100
C	41.960	42.820	41.890	43.023
H	7.380	7.020	7.160	7.300
O	38.770	38.740	36.540	36.200
N	8.490	9.080	9.220	9.480
P	0.750	0.810	0.850	0.830
S	1.010	0.890	0.900	0.990
Ca	0.420	0.230	0.290	0.540
Fe	0.180	0.043	0.049	0.160
Mg	0.130	0.070	0.110	0.140
Al	0.022	0.003	0.042	0.250
Mn	0.005	0.002	0.003	0.002
Co	0.001	0.000	0.000	0.000
Cu	0.150	0.008	0.019	0.098
Ni	0.011	0.011	0.075	0.055
Zn	0.093	0.010	0.016	0.034
Na	0.330	0.110	0.480	0.430
K	0.300	0.160	0.360	0.480
Formula	$C_{5.8}H_{12.2}O_{4.0}NP_{0.04}$	$C_{5.4}H_{10.8}O_{3.7}NP_{0.04}$	$C_{5.3}H_{10.9}O_{3.4}NP_{0.04}$	$C_{5.3}H_{10.8}O_{3.3}NP_{0.04}$

in the bulk fluid can result in rapid and drastic changes of the relative abundance and spatial distribution of organisms in biofilms (Fruhen et al., 1991). Zhang et al. (1995) found that heterotrophs, supported by soluble microbial products or metabolic products, could exist in nitrifying biofilms. Nitrifiers, however, have difficulty to survive in heterotrophic biofilms because they are likely to be out-competed by heterotrophs for dissolved oxygen and space. Inhibition or elimination of nitrifying populations by interspecies competition usually leads to a decline in nitrification efficiency, or even a failure of the process. Thus, an understanding of the effects of substrate N/COD ratio on the dynamic changes of microbial species in microbial granules is an important need.

Evolution of Heterotrophic Activities

To investigate the microbial activities and distributions of respective population, Yang et al. (2004a) determined the activities of heterotrophic, ammonia-oxidizing, and nitrite-oxidizing bacteria by the respective specific oxygen utilization rate $(SOUR)_H$, $(SOUR)_{NH_4}$, and $(SOUR)_{NO_2}$. Figure 8.6 shows the activities of heterotrophic populations in aerobic granules developed at different substrate N/COD ratios in the course of the reactor operation. The activity of heterotrophs in granules slightly decreased over operation time at the substrate N/COD ratios of 10/100, 20/100, and 30/100, only with the exception at the substrate N/COD ratio of 5/100 at which it remained unchanged with the operation time.

Evolution of Nitrifying Activities

The respective respirometric activities of ammonia oxidizers and nitrite oxidizers were described by the specific ammonium oxygen utilization rate $(SOUR)_{NH_4}$ and the specific nitritation oxygen utilization rate $(SOUR)_{NO_2}$. The activity of both ammonia oxidizers and nitrite oxidizers tended to increase with the operation time (Yang et al., 2004a). It is a reasonable consideration that the sum of $(SOUR)_{NH_4}$ and $(SOUR)_{NO_2}$, namely $(SOUR)_N$ may represent the overall activity of nitrifying populations in microbial granules. Figure 8.7 shows the values of $(SOUR)_N$ at different operation time. As can be seen, after 86 days of operation, the overall activity of

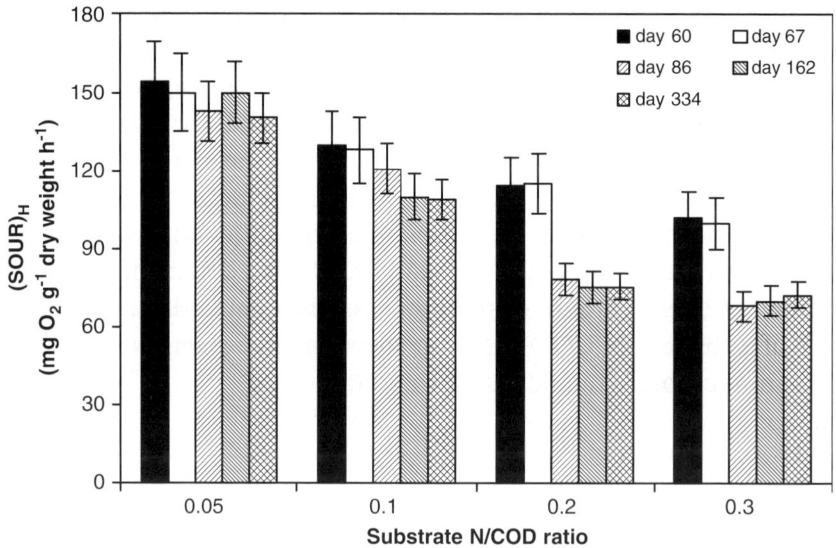

Fig. 8.6. Respirometric activities of heterotrophs in aerobic granules (Yang et al., 2004a).

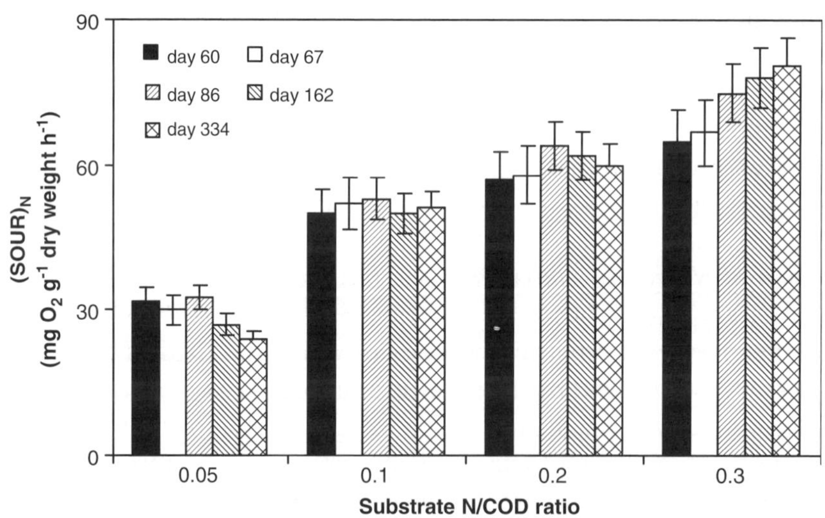

Fig. 8.7. Respirometric activities of nitrifiers in aerobic granules (Yang et al., 2004a).

nitrifying populations was approaching respective stable levels at different N/COD ratios.

These results imply that the substrate N/COD ratio would have remarkable effects on the activity distribution of ammonium-oxidizing and nitrite-oxidizing bacteria in the microbial granules, i.e. both the ammonium-oxidizing and nitrite-oxidizing activities were significantly increased with the increase of the substrate N/COD ratio, while the heterotrophic activity in the aerobic granules decreased. At high substrate N/COD ratios, heterotrophs became much less dominant, whereas nitrifying populations would be able to compete with heterotrophs, and became an important component of the aerobic granules (Figs 8.6 and 8.7). Similar phenomenon was also observed in biofilm culture (Moreau et al., 1994; Ohashi et al., 1995; Ochoa et al., 2002).

Interactions between Heterotrophic and Nitrifying Populations

The fraction of active biomass in a culture would be proportionally related to the respirometric activity (Ochoa et al., 2002). Thus, the relative abundance of nitrifying populations over heterotrophic populations can be proportionally represented by $(SOUR)_N/(SOUR)_H$. The relative abundance of nitrifying populations over heterotrophic populations in the aerobic granules can be accordingly calculated with activities values obtained (Yang et al., 2004a). The value of $(SOUR)_N/(SOUR)_H$ in the SBR run at the substrate N/COD ratio of 30/100 was 0.6 on day 60 and further increased to about 1.1 on day 86 onwards. Interactions between heterotrophic and nitrifying populations in the SBRs operated at the respective substrate N/COD ratio of 10/100 and 20/100 followed the similar pattern, i.e. $(SOUR)_N/(SOUR)_H$ gradually stabilized at a certain level. However, $(SOUR)_N/(SOUR)_H$ in the SBR operated at the substrate N/COD ratio of 5/100 almost remained constant. These seem to imply that a balance between two populations could be finally achieved in aerobic granules.

Nitrifying populations are commonly found in activated sludge and biofilms, while their quantity is generally insufficient because they would be out-competed by heterotrophs (Moreau et al., 1994). It appears from Figs 8.6 and 8.7 that nitrifying populations continued to build up over heterotrophic population in the aerobic granules until a balance between heterotrophic and nitrifying populations was reached on day 86 onwards. Aerobic granules appear to provide a protective matrix for nitrifying bacteria to grow peacefully without the risk of being washed out from the system. It may be expected that aerobic granule-based compact and efficient bioreactor for simultaneous organic removal and nitrification would be developed in near future.

Organics Removal and Nitrification

The feasibility of simultaneous removal of organics and nitrogen by microbial granules was investigated by Yang et al. (2003). Figure 8.8 shows the profiles of COD concentration and nitrification in the reactor operated at various substrate N/COD ratios. The salient points of the data are that (i) almost all influent COD is removed in the first hour; (ii) no nitrite and nitrate are produced in the reactor run at a substrate N/COD ratio of 5/100, while typical nitrification profiles were observed in the reactors operated at a respective substrate N/COD ratio of 10/100, 20/100, and 30/100; (iii) a complete nitrification occurred after the COD removal; (iv) ammonium–nitrogen removal in the first hour of the cycle was the result of microbial growth requirement for nitrogen source instead of nitrification because neither nitrite or nitrate was produced in this period; (v) no lag nitrate production with respect to nitrite formation was observed.

Basically, nitrification is completed by two kinds of bacteria, namely ammonia oxidizers responsible for nitrite formation, and nitrite oxidizers for converting nitrite to nitrate. The biological oxidation sequence can be simplified to two consecutive reactions:

$$2NH_4^+ + 3O_2 \longrightarrow 2NO_2^- + 4H^+ + 2H_2O \quad (8.1)$$

$$2NO_2^- + O_2 \longrightarrow 2NO_3^- \quad (8.2)$$

It should be realized that nitrite is an intermediate of nitrification process. The complete nitrification observed in Fig. 8.8 indicates that both ammonia oxidizer and nitrite oxidizer present sufficiently in the aerobic granules. According to Liu and Tay (2001), at least three factors would influence the nitrification profiles: (i) the relative specific growth rate of ammonia oxidizer and nitrite oxidizer in the aerobic granules; (ii) the relative ratio

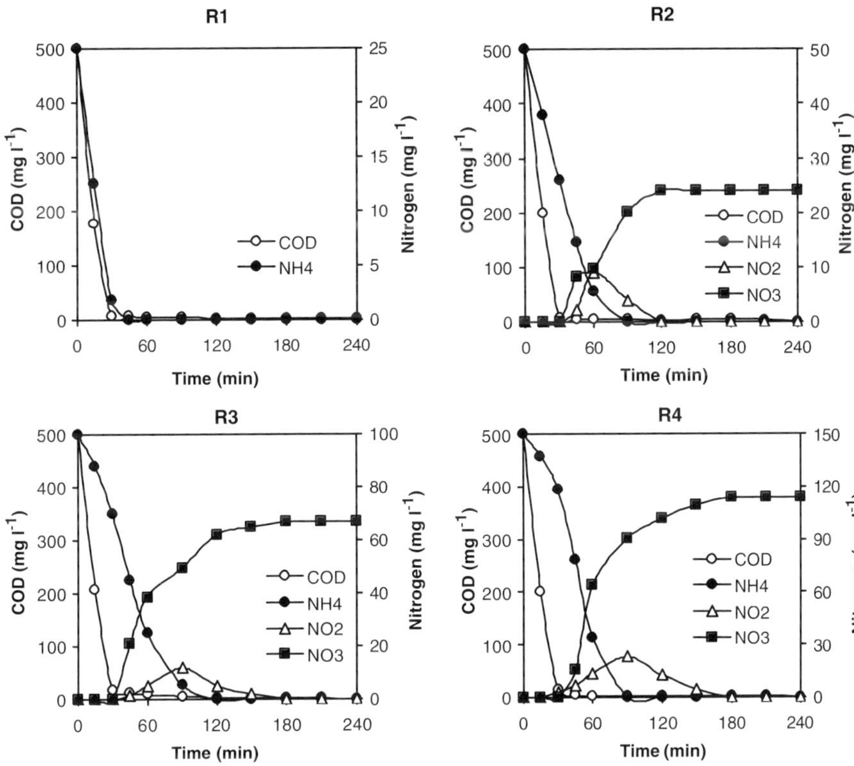

Fig. 8.8. COD and nitrogen profiles in microbial granular sludge SBRs operated at the substrate N/COD ratios of 5/100–30/100 (Yang et al., 2003).

between ammonia oxidizers and nitrite oxidizers in the aerobic granules; and (iii) the level of free ammonia. It appears that the ammonia oxidation rate is proportionally related to the substrate N/COD ratios, i.e. the higher the substrate N/COD ratio, faster is the fall of ammonium-N concentration.

Nitrogen Removal under Alternating Aerobic–Anaerobic Conditions

Microbial granules for carbon oxidation and nitrogen removal can also be developed in SBRs operated under alternating aerobic–anaerobic conditions (Qin et al., 2004b, 2005). The microbial granules cultivated under alternating aerobic–anaerobic conditions even exhibited a better settleability than those developed under pure aerobic conditions. It was found that over 95% of influent COD was removed within the first half hour of aerobic phase. Along with the COD elimination, the NH_4^+-N concentration decreased slowly to satisfy microbial growth requirement on nitrogen. After depletion of most COD, the NH_4^+-N concentration declined in a fast way, which was mainly due to nitrification. Ammonia was entirely converted to nitrate at the end of the aerobic phase. The stoichiometric amount of ethanol required for nitrate reduction can be derived from the following equation:

$$97NO_3^- + 50C_2H_5OH \longrightarrow 5C_5H_7O_2N + 75CO_2 + 84H_2O + 46N_2 + 97OH^- \quad (8.3)$$

In order to promote denitrification, external carbon source was added according to the above equation in the beginning of the anaerobic phase, and a complete denitrification was achieved in alternating aerobic–anaerobic microbial granular sludge SBRs (Qin et al., 2005).

Improved Stability of Aerobic Granules by Selecting Slow-growing Bacteria

Since aerobic bacteria grows much faster than anaerobic bacteria, the stability of aerobic granules appears to be poorer than that of anaerobic granules developed in upflow anaerobic sludge blanket (UASB) reactor. Obviously, the poor stability of aerobic granules would limit its application in wastewater treatment practice. Existing evidence shows that the stability of biofilms is closely related to the growth rate of bacteria, i.e. a higher growth rate of bacteria resulted in a weaker structure of biofilms (Tijhuis et al., 1995; Kwok et al., 1998; Liu et al., 2003b). Therefore, research attention has been given to microbial selection-based strategy for improving the stability of aerobic granules (de Kreuk and van Loosdrecht, 2004; Liu et al., 2004c). These would be very useful for the development of full-scale aerobic granules-based bioreactor for wastewater treatment.

The growth of aerobic granules after the initial cell-to-cell attachment is the net result of interaction between bacterial growth and detachment, while the balance between growth and detachment processes in turn leads to an equilibrium or stable granule size (Liu and Tay, 2002). Thus, size evolution of the microbial aggregates can be used to describe the growth of granular sludge. The specific growth rate (μ_d) by size of microbial aggregates can be defined as

$$\mu_d = \frac{dD/dt}{D} \tag{8.4}$$

in which D is the mean size of microbial aggregates, and t is the operation time. Integrating equation (8.4) gives

$$\ln D = \mu_d t + \text{constant} \tag{8.5}$$

Thus the size-dependent specific growth rate of microbial aggregate can be determined from the slope of the straight line described by equation (8.5). Figure 8.9 shows the effect of substrate N/COD ratio on μ_d (Yang et al., 2004b). It is obvious that a higher substrate N/COD ratio had resulted in a lower specific growth rate of aerobic granules with smaller size (Fig. 8.9).

Moreau et al. (1994) reported that the activity distribution of nitrifying population over heterotrophic population in biofilms was proportionally

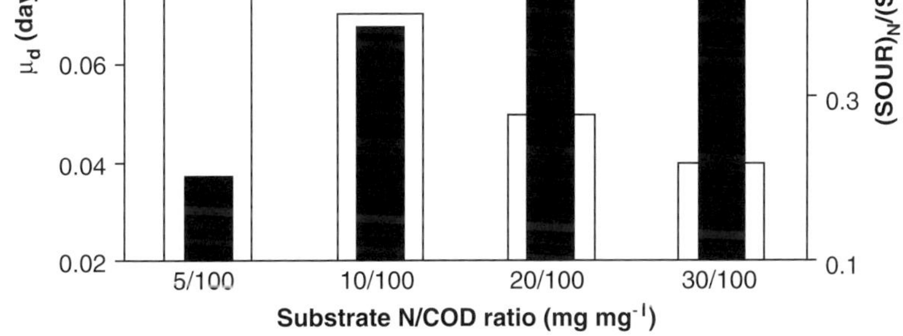

Fig. 8.9. Effect of substrate N/COD ratio on μ_d (□) and $(SOUR)_N/(SOUR)_H$ (■) of aerobic granules (Yang et al., 2004b).

related to the relative abundance of two populations under given conditions. As discussed earlier, nitrifying population in aerobic granules is enriched significantly with increasing substrate N/COD ratio. The observed growth rate and mean size at steady state aerobic granules are closely related to the substrate N/COD ratio, i.e. higher substrate N/COD ratio results in smaller granules with lower growth rate (Fig. 8.9). As nitrifying population enriched at high substrate N/COD ratio, heterotrophs in aerobic granules became less dominant. It seems that the high substrate N/COD ratio is an important factor in microbial selection with a predominantly nitrifying population.

It has been known that nitrifying bacteria grows much slower than heterotrophs, while the physical structure of nitrifying biofilms was much stronger than that heterotrophic biofilms (Oga et al., 1991). Increasing evidence shows that the observed growth rate of aerobic granules can be significantly lowered by enriching slow-growing nitrifying population, and this can be achieved through proper control of substrate N/COD or P/COD ratio (de Kreuk and van Loosdrecht, 2004; Liu et al., 2004c).

Microbial Granules for Phosphorus Removal

To eliminate phosphorus from wastewater, the phosphorus-accumulating granules (PAG) were developed at different substrate P/COD ratios in the range of 1/100 – 10/100 by weight in SBRs (Lin et al., 2003; Liu et al., 2005). Results showed that granules had typical phosphorus accumulating characteristics, with concomitant uptake of soluble organic carbon and the release of phosphate in the anaerobic stage, followed by rapid phosphate uptake in the aerobic stage.

Formation of PAGs

After a two-month operation, PAGs with a respective mean size of 1.65, 1.22, 1.03, 0.69, and 0.42 mm were formed and dominated in the SBRs run at the substrate P/COD ratios of 1/100–10/100. In contrast to the seed sludge with a very loose and irregular structure, the PAGs cultivated show a compact structure and clear spherical outer shape (Fig. 8.10).

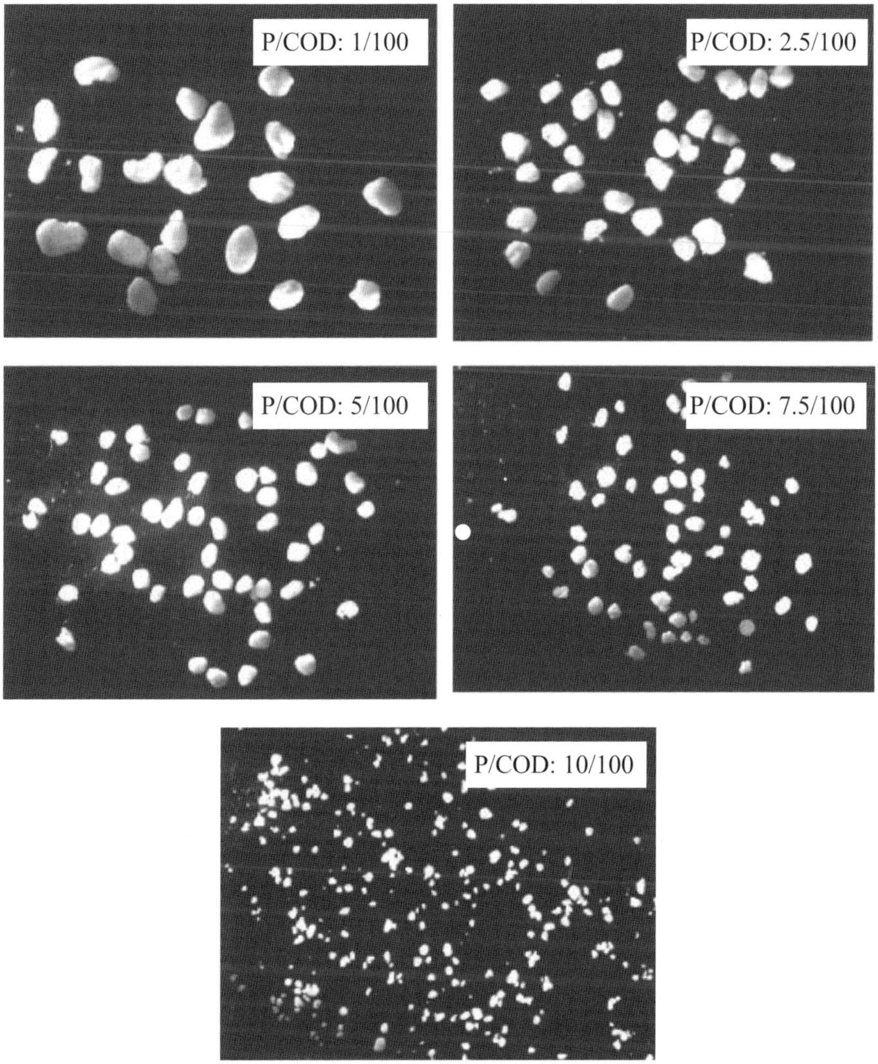

Fig. 8.10. Morphology of microbial granules developed at different substrate P/COD ratios. Bar: 2 mm (Lin et al., 2003).

Characteristics of PAGs

Settleability

The SVI of PAGs decreased with the increase in the substrate P/COD ratio. The lowest SVI of 12 ml/g was obtained at the highest substrate

P/COD ratio of 10/100 (Lin et al., 2003). This seems to imply that the substrate P/COD ratio has a profound effect on the structure of PAGs, i.e. the structural compactness of microbial granules are likely related with the feeding P/COD ratios. Compared to the seed sludge, with an SVI of 270 ml/g, the settleability of PAGs was improved distinctly. The reported SVI of non-P-accumulating aerobic granules fell in the range of 50–100 ml/g (Beun et al., 1999; Tay et al., 2001; Moy et al., 2002).

Specific Gravity

The specific gravity of PAGs appears to be proportionally related to the substrate P/COD ratio (Lin et al., 2003). This is consistent with the SVI trend obtained at different substrate P/COD ratios. In the field of environmental engineering, specific gravity has been commonly used to describe how compact or dense a microbial community is. The PAGs cultivated exhibit much denser and more compact structure as compared to the seed sludge (Lin et al., 2003).

Phosphorus Content

In EBPR processes, P in wastewater is essentially accumulated in the form of polyphosphate (poly-P) in microorganisms. The P content in PAGs was in the range of 1.9–9.3% by weight (Lin et al., 2003). If the seed sludge with a P content of 0.85% by weight is taken as reference, the P content of PAGs was 2–11 times higher than that in the seed sludge. It should be pointed out that the P content was typically in the range of 1.5–2% by dry weight in conventional activated sludge processes without the EBPR (Droste, 1997), and 4–15% in the EBPR processes (Crocetti et al., 2000; Panswad et al., 2003). These results seem to indicate PAGs developed under alternative anaerobic and aerobic conditions have a high P-accumulating ability which is essential and desired in EBPR process.

Profiles of Soluble COD and P in One Typical Cycle

Figure 8.11 shows a typical soluble COD and phosphorus profiles in one cycle operation of the microbial granular sludge SBR operated at the

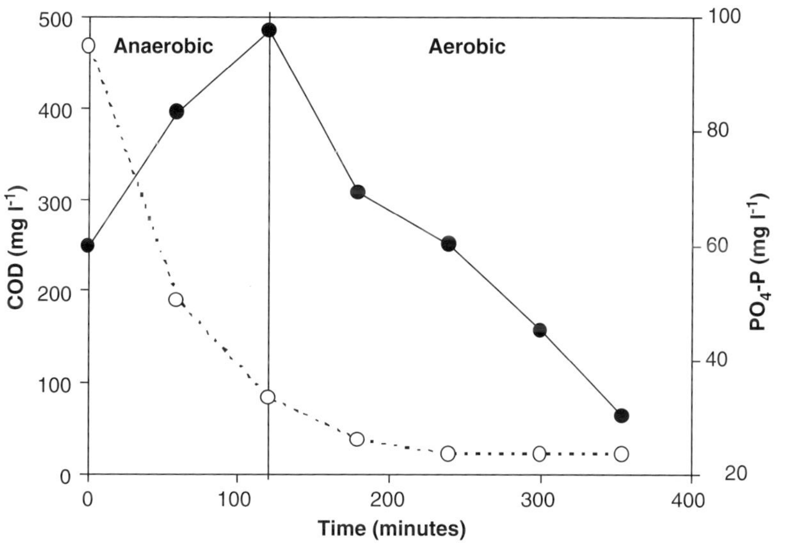

Fig. 8.11. Cycle profiles of solution COD (○) and phosphorus (●) concentrations in microbial granular sludge SBR run at substrate P/COD ratio of 10/100 (Lin et al., 2003).

substrate P/COD ratio of 10/100 (Lin et al., 2003). These results indicate that the PAGs would release phosphate with concomitant uptake of organic carbon in the anaerobic stage, and assimilate phosphate in the aerobic stage. Figure 8.11 seems to suggest that the microbial granules developed in this study have a great ability to accumulate phosphorus. In fact, Fig. 8.11 exhibits a typical biological phosphorus removal process by PAGs and the curves of COD removal and P release in anaerobic phase and uptake in aerobic phase are quite similar to those obtained in conventional EBPR processes using suspended activated sludge (Comeau et al., 1986; Wentzel et al., 1988; Hiraishi et al., 1989; Jeon and Park, 2000). Thus, it is likely that the mechanisms of the P accumulation in the aerobic granules are similar to those of P accumulation in suspended culture.

Summary

Aerobic granules for organic carbon and nitrogen removal can be successfully developed at different substrate N/COD ratios in sequencing batch reactors. Complete and efficient organics and nitrogen removal can be achieved in single granules-based SBR if the operating conditions could be properly controlled, and the system stability and nitrogen conversion capacity could be maintained in the granules-based SBR.

The phosphorus-accumulating microbial granules can also be developed at different substrate P/COD ratios in SBRs. Results showed that the PAGs had typical phosphorus accumulating characteristics and the P uptake by granules fell into a range of 1.9–9.3% by weight, which is comparable with those obtained in conventional EBPR processes. In summary, the results presented in this chapter open a door for environmental engineers to further develop a novel granules-based biological process for nutrient removal from wastewater.

References

Ballinger, S.J., Head, I.M., Curtis, T.P., & Godley, A.R. (2002). The effect of C/N ratio on ammonia oxidizing bacteria community structure in a laboratory nitrification-denitrification reactor. *Water Sci. Technol.*, *46*, 543–550.

Beun, J.J., Hendriks, A., van Loosdrecht, M.C.M., Morgenroth, E., Wilderer, P.A., & Heijnen, J.J. (1999). Aerobic granulation in a sequencing batch reactor. *Water Res.*, *33*, 2283–2290.

Beun, J.J., Heijnen, J.J., & van Loosdrecht, M.C.M. (2001). N-Removal in a granular sludge sequencing batch airlift reactor. *Biotechnol. Bioeng.*, *75*, 82–92.

Bossier, P., & Verstraete, W. (1996). Triggers for microbial aggregation in activated sludge. *Appl. Microbiol. Biotechnol.*, *45*, 1–6.

Comeau, Y., Hall, K.J., Hancock, R.E.W., & Oldham, W.K. (1986). Biochemical model for biological enhanced phosphorus removal. *Water Res.*, *20*, 1511–1521.

Crocetti, G.R., Hugenholtz, P., Bond, P.L., Schuler, A., Keller, J., Jenkins, D., & Blackall, L.L. (2000). Identification of polyphosphate accumulating organisms and design of 16S rRNA-directed probes for their detection and quantitation. *Appl. Environ. Microbiol.*, *66*, 1175–1182.

de Kreuk, M.K, & van Loosdrecht, M.C.M. (2004). Selection of slow growing organisms as a means for improving aerobic granular sludge stability. *Water Sci. Technol.*, *49*, 9–17.

Droste, R.L. (1997). *Theory and Practice of Water and Wastewater Treatment*. Wiley, New York.

Duboc, P.H., Schill, N., Menoud, L., van Gulik, W., & von Stockar, U. (1995). Measurement of sulphur, phosphorus and other ions in microbial biomass: influence on correct determination of elemental composition and degree of reduction. *J. Biotechnol.*, *43*, 145–158.

Durmaz, B., & Sanin, F.D. (2001). Effect of carbon to nitrogen ratio on the composition of microbial extracellular polymers in activated sludge. *Water Sci. Technol.*, *44*, 221–229.

Etterer, T., & Wilderer, P.A. (2001). Generation and properties of aerobic granular sludge. *Water Sci. Technol.*, *43*, 19–26.

Fletcher, M., & Floodgate, G.D. (1973). An electron-microscopic demonstration of an acid polysaccharide involved in the adhesion of a marine bacterium on solid surface. *J. Gen. Microbiol.*, *74*, 325–334.

Fruhen, M., Christan, E., Gujer, W., & Wanner, O. (1991). Significance of spatial distribution of microbial species in mixed culture biofilms. *Water Sci. Technol.*, *23*, 1365–1374.

Giokas, D.L., Daigger, G.T., von Sperling, M., Kim, Y., & Paraskevas, P.A. (2003). Comparison and evaluation of empirical zone settling velocity parameters based on sludge volume index using a unified settling characteristics database. *Water Res.*, *37*, 3821–3836.

Heldal, M., Norland, S., Fagerbakke, K.M., Thingstad, F., & Bratbak, G. (1996). The elemental composition of bacteria: a signature of growth conditions? *Mar. Pollut. Bull.*, *33*, 3–9.

Hiraishi, A., Ueda, Y., & Ishihara, J. (1989). Characterization of the bacterial population structure in an anaerobic-aerobic activated sludge system on the basis of respiratory quinine profiles. *Appl. Environ. Microbiol.*, *55*, 897–901.

Jeon, C.O., & Park, J.M. (2000). Enhanced biological phosphorus removal in a sequencing batch reactor supplied with glucose as sole carbon source. *Water Res.*, *34*, 2160–2170.

Kim, I.S., Stabnikova, E.V., & Ivanov, V.N. (2000). Hydrophobic interactions within biofilms of nitrifying and denitrifying bacteria in biofilters. *Bioprocess. Eng.*, *22*, 285–290.

Kwok, W.K., Picioreanu, C., Ong, S.L., van Loosdrecht, M.C.M., Ng, W.J., & Heijnen, J.J. (1998). Influence of biomass production and detachment forces on biofilm structures in a biofilm airlift suspension reactor. *Biotechnol. Bioeng.*, *58*, 400–407.

Laudelout, H., Simonart, P.C., & Van, P.D. (1968). Calorimetric measurement of free energy utilization by *Nitrosomonas* and *Nitrobacter*. *Archiv. Mikrobiol.*, *63*, 256–277.

Lin Y.M., Liu, Y., & Tay, J.-H. (2003). Development and characteristics of P-accumulating microbial granules in sequencing batch reactors. *Appl. Microbiol. Biotechnol.*, *62*, 430–435.

Liu, Y., & Tay, J.-H. (2001). Factors affecting nitrite build-up in nitrifying biofilm reactor. *J. Environ. Sci. Health*, *A36*, 1027–1040.

Liu, Y., & Tay, J.-H. (2002). The essential role of hydrodynamic shear force in the formation of biofilm and granular sludge. *Water Res.*, *36*, 1653–1665.

Liu, Y., Yang, S.F., & Tay, J.-H. (2003a). Elemental compositions and characteristics of aerobic granules cultivated at different substrate N/C ratios. *Appl. Microbiol. Biotechnol.*, *61*, 556–561.

Liu, Y., Lin, Y.M., Yang, S.F., & Tay, J.-H. (2003b). A balanced model for biofilms developed at different growth and detachment forces. *Process Biochemistry*, *38*, 1761–1765.

Liu, Y., Yang, S.F., Tay, J.-H., Liu, Q.S., Qin, L., & Li, Y. (2004a). Cell hydrophobicity is a triggering force of biogranulation. *Enzyme and Microbial Technology*, *34*, 371–379.

Liu, Q.Y., Liu, Y., & Tay, J.-H. (2004b). The effects of extracellular polymeric substances on the formation and stability of biogranules. *Appl. Microbiol. Biotechnol.*, *65*, 143–148.

Liu, Y., Yang, S.F., & Tay, J.-H. (2004c). Improved stability of aerobic granules by selecting slow-growing nitrifying bacteria. *J. Biotechnol.*, *108*, 161–169.

Liu, Y., Lin, Y.M., & Tay, J.-H. (2005). The elemental compositions of P-accumulating microbial granules developed in sequencing batch reactors. *Process Biochem.*, *40*, 3258–3262.

Lopes, F.A., Vieira, M.J., & Melo, L.F. (2000). Chemical composition and activity of a biofilm during the start-up of an airlift reactor. *Water Sci. Technol.*, *41*, 105–111.

Mahoney, E.M., Varangu, L.K., Cairns, W.L., Kosaric, N., & Murray, R.G.E. (1987). The effect of calcium on microbial aggregation during UASB reactor start-up. *Water Sci. Technol.*, *19*, 249–260.

Moreau, M., Liu, Y., Capdeville, B., Audic, J.M., & Calvez, L. (1994). Kinetic behaviors of heterotrophic and autotrophic biofilm in wastewater treatment processes. *Water Sci. Technol.*, *29*, 385–391.

Moy, B.Y.P., Tay, J.-H., Toh, S.K., Liu, Y., & Tay, S.T.-L. (2002). High organic loading influences the physical characteristics of aerobic granules. *Lett. Appl. Microbiol.*, *34*, 407–412.

Ochoa, J.C., Colprim, J., Palacios, B., Paul, E., & Chatellier, P. (2002). Active heterotrophic and autotrophic biomass distribution between fixed and suspended systems in a hybrid biological reactor. *Water Sci. Technol.*, *46*, 397–404.

Oga, T., Suthersan, S., & Ganczarczyk, J.J. (1991). Some properties of aerobic biofilms. *Environ. Technol.*, *12*, 431–440.

Ohashi, A., Viraj de Silva, D.G., Mobarry, B., Manem, J.A., Stahl, D.A., & Rittmann, B.E. (1995). Influence of substrate C/N ratio on the structure of multi-species biofilms consisting of nitrifiers and heterotrophs. *Water Sci. Technol.*, *32*, 75–84.

Panswad, T., Doungchai, A., & Anotai, J. (2003). Temperature effect on microbial community of enhanced biological phosphorus removal system. *Water Res.*, *37*, 409–415.

Peng, D., Bernet, N., Delgenes, J.P., & Moletta, R. (1999). Aerobic granular sludge – a case study. *Water Res.*, *33*, 890–893.

Pitryuk, A.V., Pusheva, M.A., & Sorokin, V.V. (2002). Elemental composition of extremely alkaliphilic anaerobic bacteria. *Microbiol.*, *71*, 24–30.

Prinčič A., Mahne, I., Megušar, F., Aaul, E.A., & Tiedje, J.M. (1998). Effects of pH and oxygen and ammonium concentrations on the community structure of nitrifying bacteria from wastewater. *Appl. Environ. Microbiol.*, *64*, 3584–3590.

Qin, L., Liu, Y., & Tay, J.-H. (2004a). Effect of settling time on aerobic granulation in sequencing batch reactor. *Biochem. Eng. J.*, *21*, 47–52.

Qin, L., Tay, J.-H., Yang, S.F., & Liu, Y. (2004b). Aerobic granulation under alternating aerobic and anaerobic conditions in sequencing batch reactors. *Water Environmental Management Book Series* (eds. Lens, P., & Stuetz, R.), IWA Publishing, London, UK, 3–10.

Qin, L., Liu, Y., & Tay, J.-H. (2005). Denitrification on poly-β-hydroxybutyrate in microbial granular sludge sequencing batch reactor. *Water Res.*, *39*, 1503–510.

Robinson, J.A., Trulear, M.G., & Characklis, W.G. (1984). Cellular reproduction and extracellular polymer formation by *Pseudomonas aeruginosa* in continuous culture. *Biotechnol. Bioeng.*, *26*, 1409–1417.

Rouxhet, P.G., & Mozes, N. (1990). Physical chemistry of the interaction between attached microorganisms and their support. *Water Sci. Technol.*, *22*, 1–16.

Schmidt, J.E., & Ahring, B.K. (1996). Granular sludge formation in upflow anaerobic sludge blanket (UASB) reactors. *Biotechnol. Bioeng.*, *49*, 229–246.

Sousa, M., Azeredo, J., Feijo, J., & Oliveira, R. (1997). Polymeric supports for the adhesion of a consortium of autotrophic nitrifying bacteria. *Biotechnol. Tech.*, *11*, 751–754.

Sutherlan, I.W. (2001). Biofilm exopolysaccharides: a strong and sticky framework. *Microbiol.*, *147*, 3–9.

Tay, J.-H., Xu, H.L., & Teo, K.C. (2000). Molecular mechanism of granulation. I: H^+ translocation-dehydration theory. *J. Environ. Eng.*, *126*, 403–410.

Tay, J.-H., Liu, Q.S., & Liu, Y. (2001). The effects of shear force on the formation, structure and metabolism of aerobic granules. *Appl. Microbiol. Biotechnol.*, *57*, 227–233.

Tijhuis, L., van Loosdrecht, M.C.M., & Heijnen, J.J. (1995). Dynamics of biofilm detachment. *Biotechnol. Bioeng.*, *45*, 481–487.

Tsuneda, S., Park, S., Hayashi, H., Jung, J., & Hirata, A. (2001). Enhancement of nitrifying biofilm formation using selected EPS produced by heterotrophic bacteria. *Water Sci. Technol.*, *43*, 197–204.

Tsuneda, S., Aikawa, H., Hayashi, H., Yuasa, A., & Hirata, A. (2003). Extracellular polymeric substances responsible for bacterial adhesion onto solid surface. *FEMS Microbiol. Lett.*, *223*, 287–292.

Vandevivere, P., & Kirchman, D.L. (1993). Attachment stimulates exopolysaccharide synthesis by bacteria. *Appl. Environ. Microbiol.*, *59*, 3280–3286.

Vrede, K., Heldal, M., Norland, S., & Bratbak, G. (2002). Elemental composition (C, N, P) and cell volume of exponentially growing and nutrient-limited bacterioplankton. *Appl. Environ. Microbiol.*, *68*, 2965–2971.

Wang, Z.W., Liu, Y. & Tay, J.-H. (2005). Distribution of EPS and cell surface hydrophobicity in aerobic granules. *Appl. Microbiol. Biotechnol.*, *69* (4), 469–473.

Wentzel, M.C., Loewenthal, R.E., Ekama, G.A., & Marais, G.V.R. (1988). Enhanced polyphosphate organism cultures in activated sludge system. *Water SA*, *14*, 81–92.

Wuertz, S., Pfleiderer, P., Kriebitzsch, K., Griebe, T., Coello-Oviedo, D., Wilderer, P.A., & Flemming, H.C. (1998). Extracellular redox activity in activated sludge. *Water Sci. Technol.*, *37*, 379–384.

Yang, S.F., Tay, J.-H., & Liu, Y. (2003). A novel granular sludge sequencing-batch reactor for organic and nitrogen removal from wastewater. *J. Biotechnol.*, *106*, 77–86.

Yang, S.F., Tay, J.-H., & Liu, Y. (2004a). Respirometric activities of heterotrophic and nitrifying populations in aerobic granules developed at different substrate N/COD ratios. *Current Microbiol.*, *49*, 42–46.

Yang, S.F., Liu, Q.S., Tay, J.-H., & Liu, Y. (2004b). Growth kinetics of aerobic granules developed in sequencing batch reactors. *Lett. Appl. Microbiol.*, *38*, 106–112.

Yang, S.F., Tay, J.-H., & Liu, Y. (2005). Effect of substrate N/COD ratio on the formation of aerobic granules. *J. Environ. Eng.*, *131*, 86–92.

Yu, H.Q., Tay, J.-H., & Fang, H.H.P. (2001). The role of calcium in sludge granulation during UASB reactor start-up. *Water Res.*, *35*, 1052–1060.

Zhang, T.C., Fu, T.C., & Bishop, P.L. (1995). Competition for substrate and space in biofilms. *Water Environ. Res.*, *67*, 992–1003.

Zita, A., & Hermansson, M. (1997). Effects of bacterial cell surface structures and hydrophobicity on attachment to activated sludge flocs. *Appl. Environ. Microbiol.*, *63*, 1168–1170.

Chapter 9

Removal of Phenol from Wastewater by Microbial Granules

Stephen Tiong-Lee Tay

Sources and Applications of Phenol

Phenol, or hydroxybenzene, was first discovered in 1834 and used in the raw state as creosote to prevent the weathering of railway ties and ships' timber, as well as to reduce odors in decomposed sewage. Phenol is derived from coal tar distillation and is manufactured by synthetic processes such as oxidation of methylethylbenzene, oxidation of toluene and heating of monochlorobenzene with sodium hydroxide under high pressure. In terms of production volume, phenol ranks among the top 40 chemicals in the United States, with 4.77 billion pounds of phenol manufactured in 1998 (van Schie and Young, 2000). Phenol is an important industrial chemical, and is used in the production of polycarbonate resins, explosives, paints, inks, perfumes, wood preservatives (as pentachlorophenol), textiles, drugs, and as an antibacterial and antifungal agent. Phenol is also used in medicine as a topical anesthetic or antiseptic.

Phenol concentrations from natural sources are typically much lower than from anthropogenic activities. Any type of plant material, including leaves, shoots, roots, and flowers, can potentially leach phenolic compounds. Some of these compounds, such as 4-hydroxybenzoate, serve as chemo-attractants for soil microbes, as recognition signals for plant–microbe interactions, or as transcription signals for nodulation genes in *Rhizobium* (Kape et al., 1991). Natural phenols may also be found in marine plants and algae (Boyd and Carlucci, 1993), or be excreted in the feces and urine of animals whose intestines are inhabited by microorganisms that feed on aromatic amino acid precursors such as tyrosine (Smith and Macfarlane, 1997).

Contamination of Environment with Phenol

Due to its widespread use and its ubiquity, phenol is a major pollutant of the environment. Industries that produce or use phenol may at some point release this compound into the environment. It is therefore not surprising to find that phenol is a commonly found waste by-product in many industries, including petroleum refining, petrochemical, coke conversion, pharmaceutical, and resin manufacturing plants. Phenol concentrations of up to $10{,}000\,mg\,l^{-1}$ have been reported in industrial wastewaters (Fedorak and Hrudey, 1988). Without proper treatment, industrial wastewaters would become potentially important sources of anthropogenic phenol into the environment.

Phenol and its structurally related compounds are toxic at relatively low concentrations and are listed as priority pollutants by the US Environmental Protection Agency (Ghisalba, 1983). Phenol can be toxic to some aquatic species at concentrations in the low $mg\,l^{-1}$ range (Brown et al., 1967) and causes taste and odor problems in drinking water at far lower concentrations (Rittmann and McCarty, 2001). Hence the removal of phenol from wastewater is of obvious interest.

Phenol can be removed by solvent extraction, adsorption, chemical oxidation, incineration, and other non-biological treatment methods, but these methods suffer from serious drawbacks such as high cost and formation of hazardous by-products (Loh et al., 2000). Biological degradation is generally preferred due to lower costs and the possibility of complete mineralization. However, phenol-containing wastewater is difficult to treat because of substrate inhibition, whereby microbial growth and concomitant biodegradation of phenol are hindered by the toxicity exerted by high concentrations of the substrate itself.

Microbial Resistance to Phenol Toxicity

Phenol exerts a general bactericidal effect because of the compound's ability to partition into cell membranes, which leads to a loss of cytoplasmic membrane integrity. Phenol toxicity results in disruption of microbial activities associated with energy transformations, membrane barrier functions, and membrane protein functions, and causes eventual cell death. Nevertheless, microorganisms are known to develop mechanisms to resist and survive phenol at concentrations that are normally inhibitory to microbial activity. These mechanisms include isomerization of *cis*-unsaturated fatty acids to the *trans*-configuration and increase in proportion of saturated fatty acids to unsaturated fatty acids. Such adaptive responses to phenol exposure allow for chains of fatty acid molecules to be more closely aligned to improve the structural rigidity of cell membranes, thus compensating for the increased membrane fluidity induced by phenol partitioning (Heipieper et al., 1991; Keweloh et al., 1991; Yap et al., 1999). In fact, several mechanisms that decrease membrane fluidity in response to substrate toxicity have been proposed for *Pseudomonas putida* (Heipieper et al., 1992), *Escherichia coli* (Keweloh et al., 1991), and *Vibrio* species (Okuyama et al., 1991). Bacteria that possess such resistance mechanisms to counteract high concentrations of phenol would therefore be of considerable practical interest for deployment in biodegradation processes where phenolic compounds can exert a toxic or inhibitory effect.

Aerobic Biodegradation of Phenol

In view of the widespread occurrence of phenol, microorganisms can be found in many environments that are able to use phenol as a carbon and energy source. These microorganisms include both aerobic and anaerobic microorganisms. Many aerobic phenol-degrading bacteria have been isolated since the beginning of the 1900s, and the pathways for phenol degradation are now well established (van Schie and Young, 2000).

Aerobic degradation of phenol by microorganisms is mainly based on the *ortho*- and *meta*-cleavage pathways (Muller and Babel, 1996). During the first step of the aerobic phenol degradation pathway, molecular oxygen is used by the phenol hydroxylase enzyme to add a second hydroxyl group in the *ortho*-position to the one already present to produce catechol, which can then be degraded by either the *ortho*- or the *meta*-cleavage pathways. In the *ortho*- or β-ketoadipate pathway, the aromatic ring is cleaved between the catechol hydroxyls by intradiol fission with a catechol 1,2-dioxygenase (C12O) enzyme. The *cis,cis*-muconate that results is metabolized via β-ketoadipate to produce the common tri-carboxylic acid cycle intermediates, succinate, and acetyl-CoA. In the *meta*-pathway, extradiol fission occurs adjacent to the two hydroxyl groups in catechol with a catechol 2,3-dioxygenase (C23O) enzyme to produce 2-hydroxymuconic semialdehyde. This compound is further metabolized to intermediates of the tricarboxylic acid cycle.

Many bacteria have been encountered that possess either one or two of the *ortho*- and *meta*-pathways. The enzymes phenol hydroxylase and C12O or C23O catalyze the first and second steps of phenol degradation, respectively. Phenol hydroxylases are generally classified as simple single-component enzymes using flavoprotein monooxygenases or multicomponent enzymes with multiple proteins. The multicomponent phenol hydroxylase is organizationally similar to the multicomponent mono- and dioxygenases involved in the degradation of toluene, benzene, naphthalene, and methane, and is considered to be the major form of phenol hydroxylase in the environment (Harayama et al., 1992). The genes necessary for the expression of phenol hydroxylase can be either chromosome- or plasmid-encoded. For instance, multicomponent phenol hydroxylase together with genes encoding the *meta*-cleavage are organized in operons located on the chromosome of *Pseudomonas sp.* (Ng et al., 1995) or on the TOL plasmid (Nordlund et al., 1990).

Anaerobic Biodegradation of Phenol

In aerobic phenol degradation, the highly reactive molecular oxygen is used for the initial attack on the aromatic ring as well as for the final ring cleavage step. A similarly reactive co-substrate is not available in anaerobic pathways. Therefore, anaerobic aromatic pathways use different mechanisms to degrade and cleave the compounds. In contrast to aerobic pathways, the reactions involved in anaerobic aromatic metabolism are largely reductive modifications of the substrate. In particular, the actual dearomatizing reaction in the different anaerobic pathways proceeds by reduction of the aromatic ring to non-aromatic cyclohexane-derivatives.

Pure cultures of anaerobically respiring bacteria can completely oxidize phenol to carbon dioxide using nitrate, ferric iron or sulfate as terminal electron acceptor (van Schie and Young, 2000). At least two phenol degradation pathways have been suggested to occur under methanogenic conditions (Karlsson et al., 2000). The most studied anaerobic pathway is phenol transformation via the "benzoyl-CoA" pathway. Not surprisingly, the efficiency of anaerobic pathways is not as high as in aerobic pathways. Hence, most bioremediation efforts directed at phenol removal invariably involve aerobic processes.

Conventional Biological Treatment of Phenol-containing Wastewater

The ability of bacteria to degrade phenol has many practical applications, such as the biological treatment of phenol-containing industrial wastewater and the bioremediation of sites polluted with phenolic compounds. The biological treatment of phenol wastewater has been mostly based on conventional continuous aerobic activated sludge systems. Activated sludge is a suspended growth process that began in England at the turn of the last century and has been widely used in municipal and industrial wastewater treatment. This process essentially consists of an aerobic treatment that oxidizes organic matter and other wastewater contaminants to carbon dioxide, water, and new cell biomass. Air is supplied by diffused or mechanical aeration and the microbial cells form activated sludge flocs that are allowed to settle in a secondary clarifier.

Although phenol removal has been carried out for many years by activated sludge systems, the treatment process has been known to break down because of the toxicity effects of high phenol concentrations encountered during episodes of fluctuations in phenol loads and of high phenol loading rates in excess of 1 kg phenol m^{-3} d^{-1} (Watanabe et al., 1996, 1999; Kibret et al., 2000). Phenol toxicity can cause inhibition of the degradation processes, decrease in the settleability and washout of sludge biomass, high phenol concentrations in the effluent, and lead to the unrecoverable failure of activated sludge systems, often rapidly.

The sequencing batch reactor (SBR) represents a promising form of biological wastewater treatment technology belonging to the group of so-called fill and draw reactors (Wilderer et al., 2001). The SBR process is a variable volume, suspended growth, biological wastewater treatment technology that is characterized by a repetitive batch cycle consisting of several successive phases (usually fill, react, settle, decant, and idle), each lasting for a defined period. Each phase can be adjusted according to its position and function within the batch cycle to satisfy specific treatment objectives. Unlike activated sludge systems, aeration and sedimentation–clarification occur sequentially in the same vessel in SBR technology. Because of its applicability to simple automation, the ease with which its operation can be modified, its single-tank design and the ability to select robust microbial communities, SBR technology has been gaining widespread acceptance within the engineering community (Mace and Mata-Alvarez, 2002). However, in comparison with continuous flow activated sludge systems, the knowledge base for SBR performance during practical situations has not been fully developed, and there are few reports on the use of SBR for treatment of phenol. Although an SBR had been recently reported to treat phenol wastewater at a high phenol loading rate of 3.1 g phenol $l^{-3}d^{-1}$, the settling ability of flocculated sludge in that reactor was generally poor, even at a low phenol loading rate of 0.52 g phenol l^{-3} d^{-1} (Yoong et al., 2000).

Use of Immobilized Cells for Phenol Biodegradation

The substrate inhibition difficulties associated with high-strength phenolic wastewaters can be overcome by strategies involving immobilization of bacterial cells (Keweloh et al., 1989). Cells that are immobilized onto various support materials are more resilient to chemical toxicity and can tolerate higher phenol concentrations than their suspended counterparts. For instance, cells of *Pseudomonas putida* immobilized in hollow fiber membranes degraded phenol at concentrations up to 3500 mg l^{-1}, albeit at relatively low rates, while their suspended counterparts encountered complete substrate inhibition at the high phenol concentrations tested (Loh et al., 2000). Cells of *Rhodococcus erythropolis* UPV-1 immobilized on diatomaceous earth demonstrated enhanced respiratory activity and a shorter lag phase preceding phenol degradation, degrading phenol at a volumetric productivity of 11.5 g phenol l^{-1} d^{-1} (Prieto et al., 2002). These immobilizations require carrier materials for biofilm attachment, necessitating higher investment and operating costs.

Recent research efforts have focused on aerobic granulation as a new form of cell immobilization for exploitation in biological wastewater treatment (Morgenroth et al., 1997; Beun et al., 1999; Tay et al., 2001). Aerobic granulation technology can overcome the disadvantages associated with the use of carrier materials in traditional cell immobilization systems. Aerobic granules are self-immobilized microbial aggregates that are cultivated in SBRs without reliance on artificial surfaces for biofilm attachment, hence rendering carrier material and settling devices unnecessary. The basis for the formation of aerobic granules in the SBR is a repetitive selection for sludge particles such that denser components are retained in the system while lighter and dispersed particles are washed out. These aerobic granules have a strong, compact microbial structure, good settling ability and high biomass retention, with the ability to handle high organic loading rates (Moy et al., 2002). The initial studies have involved cultivation of aerobic granules on simple and relatively benign substrates such as glucose and acetate, using activated sludge as inoculum. However, aerobic granules should be suitable for application in degrading toxic chemicals such as phenol, as the aggregation of microorganisms into compact aerobic granules should confer additional benefits such as protection against predation and resistance to chemical toxicity.

Cultivation of Aerobic Granules for Phenol Removal from Wastewater

The cultivation of aerobic phenol-degrading granules was first achieved by conditioning municipal activated sludge in batch culture for a period of two months by incubating with phenol which gradually increased in concentration from 50 mg l^{-1} to 500 mg l^{-1} (Jiang et al., 2002). The acclimated activated sludge was then inoculated into a column-type SBR, which was fed with a synthetic wastewater with phenol as a sole carbon source. The reactor was operated sequentially in 4 h cycles (2 min fill, 205–230 min aeration, 5–30 min settling, and 3 min effluent withdrawal) with a volumetric exchange ratio of 50% to give a hydraulic residence time (HRT) of 8 h and a phenol loading rate was 1.5 g phenol l^{-1} day^{-1}. A settling period of 30 min was initially imposed to avoid excessive washout of the acclimated biomass, and subsequently reduced to 10 min on day 20 and to

5 min from day 35 to the end of reactor operation on day 68. Granules first appeared on day 9 of reactor operation and eventually displaced the activated sludge flocs to become the dominant form of biomass within the reactor. At steady state, 80% by volume of the granules were between 0.35 and 0.60 mm in size. The step-wise decrease in settling periods selected for granules with improved settling characteristics and resulted in a concomitant increase in biomass concentration and nearly complete phenol removal in the reactor. Compared to a phenol concentration of 500 mg l^{-1} in the influent, the phenol concentration in the effluent decreased to below 0.2 mg l^{-1} beyond day 32 of reactor operation. To investigate the feasibility of using aerobically grown microbial granules for high-rate phenol biodegradation, the reactor operation was extended by imposing a higher loading rate of 2.5 kg phenol m^{-3} d^{-1} (Tay et al., 2004). Granules cultivated at this higher phenol loading did not diminish in their ability to remove phenol, and effluent phenol concentrations below 0.2 mg l^{-1} continued to be maintained.

The ability of granules to degrade phenol was evaluated by monitoring phenol disappearance at different phenol concentrations in batch experiments. The specific phenol degradation rate increased with phenol concentration from 0 to 500 mg phenol l^{-1}, peaked at 1.4 g phenol g VSS^{-1} d^{-1}, and declined with further increase in phenol concentration as substrate inhibition effects became important (Fig. 9.1). A kinetic analysis of the degradation data was performed based on Haldane's formula for self-inhibition as follows:

$$V = V_{max} S/[K_s + S + (S^2/K_i)] \tag{9.1}$$

where V and V_{max} are the specific and the theoretical maximum specific substrate degradation rates (g phenol g VSS^{-1} d^{-1}), respectively, and S, K_s, and K_i are the substrate concentration, half-saturation constant, and inhibition constant (mg phenol l^{-1}), respectively. There was a good fit between the degradation data and the Haldane equation, with calculated kinetic parameters of $V_{max} = 5.6$ g phenol g VSS^{-1} d^{-1}, $K_s = 481$ mg l^{-1}, and $K_i = 213$ mg l^{-1} (Jiang et al., 2002).

The self-immobilization of microbial cells into granules allowed the cells inside the granules to tolerate high phenol concentrations and achieve good structural properties and biodegradation performance. The cell immobilization created a diffusion barrier and established a concentration

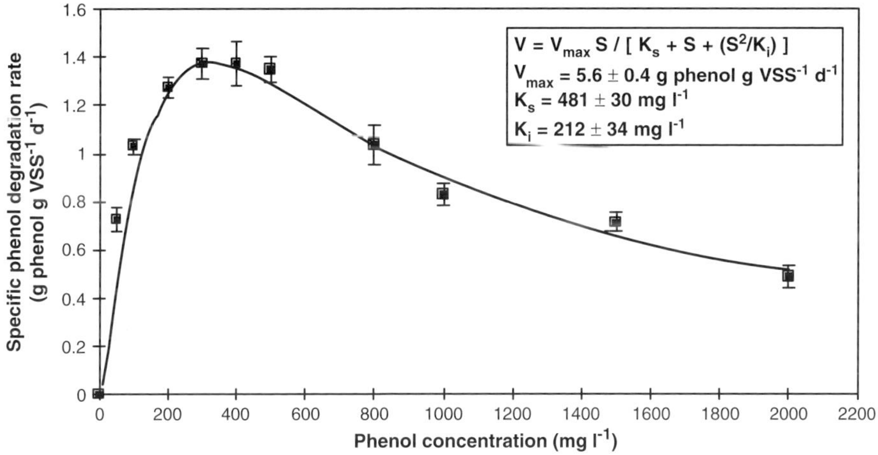

Fig. 9.1. Specific phenol degradation rates of aerobic granules at different phenol concentrations (Jiang et al., 2002).

gradient that sheltered the microbial cells beneath the protective barrier by diminishing the toxic chemical below some threshold value to allow continued microbial activity and substrate utilization (Villaverde and Fernandez-Polanco, 1999). Although the specific phenol degradation rates gradually declined at phenol concentrations higher than 500 mg l^{-1}, significant rates of phenol degradation could still be achieved up to a phenol concentration of 2000 mg l^{-1}. The ability of the granules to degrade phenol was superior to activated sludge, where the highest specific phenol degradation rates were reported at phenol concentrations below 100 mg l^{-1} (Watanabe et al., 1996).

Such application of immobilized cells in wastewater treatment permits the degradation of higher concentrations of toxic pollutants than can be achieved with free cells. Immobilized cells are better protected against phenolic and other similarly inhibitory compounds. This has been documented for many microbial systems, including alginate-encapsulated cells that tolerated higher phenol concentrations than free cells (Bastos et al., 2001), and cells entrapped in hollow-fiber membranes that mitigated the effects of phenol inhibition (Loh et al., 2000). With the protection offered by immobilization, the need for the adaptation period normally required by free cells is eliminated, and the uptake of substrate is enhanced in comparison with free cells in the bulk liquid (Diaz et al., 2002; Moslemy et al., 2002). In addition to diffusion limitation, other possible mechanisms for tolerance of elevated phenol concentrations include binding of phenol by cells on biofilm exteriors that allowed internal bacterial cells to multiply without any inhibition, and modification in lipid composition of cell membranes to compensate for the increase in membrane fluidity induced by phenol (Keweloh et al., 1989; Yap et al., 1999). Although the underlying mechanism for phenol tolerance in the granules has not been pinpointed, and may be the result of several synergistic mechanisms, it is clear from this study that the formation of dense, compact granules can facilitate growth of microorganisms in the bioreactor.

While the granular structure serves an important function in protecting microorganism within the granules against chemical toxicity, this protection may also paradoxically create problems associated with slow diffusion of nutrients and oxygen into and waste metabolites out of the granules. The pulse feeding and high phenol loading regime created a situation of high phenol concentration in the bulk liquid during the initial part of each SBR cycle, and this probably enhanced the penetration of phenol substrate into the granule interior. Moreover, the limited solubility of oxygen vis-à-vis organic substrates such as phenol means that problems that may be caused by the onset of diffusion limitation are likely to be oxygen-related (Beun et al., 2002). Previous investigations into the microstructure and ecology of aerobic granules (Tay et al., 2002, 2003) support the view that oxygen diffusion may not be limiting for small granules, and that anaerobiosis and cell death from diffusion limitation might occur in the interiors of larger granules, although the slightly looser structure of these granules would have a compensatory effect and would have allowed them to have better access to oxygen and nutrients. Still, diffusion limitation can pose a serious problem in large granules, since not all the microorganisms can actively carry out the biodegradation of target substrates. In order to exploit aerobic granulation technology for efficient treatment of high-load wastewaters, operating controls should be imposed to limit the granule size and ensure that the granules consist entirely of actively biodegrading cells.

Microbial Response of Aerobic Granules to High Phenol Loading

While overexposure to phenol is usually associated with decreases or complete loss in specific growth rate, specific oxygen utilization and enzyme activity, microorganisms in the aerobic granules should be capable of a variety of adaptive physiological responses to tolerate phenol toxicity. To investigate how phenol loading affected the structure, activity and metabolism of aerobic granules, four SBRs were fed with phenol as sole carbon and energy source at loading rates of 1.0, 1.5, 2.0, and 2.5 g phenol l^{-1} d^{-1} (Jiang et al., 2004a). After about two months of operation, all four reactors reached a steady state, as evidenced by stable biomass concentrations and constant phenol removal efficiencies. Compact granules with good settling ability were maintained at loadings up to 2.0 g phenol l^{-1} d^{-1}, but structurally weakened granules with enhanced production of extracellular polymers and proteins and significantly lower hydrophobicities were observed at the highest loading of 2.5 g phenol l^{-1} d^{-1}. Specific oxygen uptake rate, catechol 2,3-dioxygenase (C23O) and catechol 1,2-dioxygenase (C12O) activities peaked at a loading of 2.0 g phenol l^{-1} d^{-1}, and declined thereafter. The granules degraded phenol completely in all four reactors, mainly through the *meta*-cleavage pathway as C23O activities were significantly higher than C12O activities.

At the highest loading applied, the anabolism and catabolism of microorganisms were regulated such that phenol degradation proceeded exclusively via the *meta*-pathway, apparently to produce more energy for overstimulation of protein production as additional protection against phenol toxicity. Microorganisms are known to regulate synthesis of extracellular polymers (ECPs) and modify ECP properties as a microbial response against the effect of antimicrobial agents. ECPs can form a protective shield for the cells against the adverse influences of the external environment, and delay or prevent toxicants from reaching microbes by acting as a diffusion limitation barrier. Such preferential production of proteins over polysaccharides in the ECPs in the aerobic granules has also been observed in other biofilms exposed to phenol (Fang et al., 2002). Possible explanations for the elevated production of proteins include induction of heat shock-like proteins as a defense mechanism against high phenol concentrations, and induction of special proteins that could be involved in the catalytic degradation of phenol and other potentially toxic compounds (Benndorf et al., 2001).

Degradation of phenol may proceed via either the *ortho* (C12O) or the *meta* (C23O) cleavage pathway, which are often found to occur simultaneously in the same strain (Kiesel and Muller, 2002). With aerobic granules, phenol biodegradation proceeded mainly via the *meta*-pathway, as C23O activities were significantly higher than C12O activities (Jiang et al., 2004a). Previous studies have shown that the *ortho*-pathway dominated the *meta*-pathway at low growth rates due to affinity reasons, whereas the *meta*-pathway attained the highest growth rates (Filonov et al., 1997). Half-saturation constants (K_S) are also usually higher for the *meta*-pathway than for the *ortho*-pathway (Muller and Babel, 1996). Thus high-affinity/low-rate properties are found at low substrate concentrations in contrast to low-affinity/high-rate properties in situations with increased levels of substrate. From a kinetics point of view, the high phenol concentrations employed might help explain the observed predominance of the *meta* over the *ortho*-cleavage pathway in the aerobic phenol-degrading granules.

It should be noted that the choice of cleavage pathways is also mediated by metabolic factors. For kinetics reasons, the shorter route for energy production through the *meta*-pathway corresponds to a higher overall growth rate (Kiesel and Muller, 2002). This may be considered a selective advantage when alternative metabolic routes have to compete successfully for a common carbon/energy source whenever there is excess substrate, but the rate increase is obtained at the expense of a lower efficiency of carbon conversion into biomass. It is very likely that the selection pressure exerted by high phenol loads can drive the microbial community to regulate its metabolic pathways so as to maintain a balance with the external pressure by consuming non-growth-associated energy to counteract the toxicity-related inhibition of cellular activity and deterioration in granule structure. Part of non-growth-associated energy produced by metabolism might be used to maintain the integrity of cell membranes, since energy expended for this purpose would be expected to be higher at higher phenol concentrations, and part of the energy was directed towards the production of ECPs as shown earlier (Jiang et al., 2004a).

Bacterial Diversity and Functions in Aerobic Phenol-degrading Granules

Aerobic granules can be viewed as a special form of biofilm, but without carriers for biofilm attachment. Growth environments for biofilm communities are different from planktonic communities, and microbial communities in attached biofilms have been shown to be highly distinct from the suspended biomass, even within a single reactor system. Recognizing the diversity and the linkages among the key functional groups in any given biological system can lead to better ways to model and understand diversity and function as well as to improve process stability. In a recent study, culture-independent and culture-dependent methods were used in combination to study the microbial community of aerobic phenol-degrading granules and to isolate, characterize, and identify ecologically relevant microorganisms (Jiang et al., 2004b).

The direct isolation technique was used to obtain bacterial colonies by incubating biomass from aerobic phenol-degrading granules on MP medium agar plates supplemented with 500 mg phenol l^{-1}. A final set of ten strains, designated PG-01 to PG-10, was assembled after screening of 16S rRNA genes with REP-PCR (Table 9.1). Seven of the ten isolates belonged to the β- or γ-*Proteobacteria* group. These culture-based data are consistent with previous studies which demonstrated that β- and γ-*Proteobacteria* constitute a large fraction of the bacteria in wastewater treatment plants (Bond et al., 1995; Snaidr et al., 1997) or in glucose-fed aerobic granules (Tay et al., 2002). Members of β-*Proteobacteria* have also been implicated in phenol degradation in activated sludge, as demonstrated in isolation experiments (Watanabe et al., 1998).

Another interesting observation was the prevalence of gram-positive high G+C bacteria in the phenol-degrading aerobic granules. In contrast, gram-positive high G+C bacteria were not dominant members in phenol-degrading activated sludge systems (Watanabe et al., 1998, 1999; Whiteley et al., 2001). These observations could probably be explained by the fact that high G+C bacteria preferred to grow in attached biofilms than to remain in a planktonic state (Lehman et al., 2001; Tresse et al., 2002). These microorganisms are also known to be resilient to external stresses, because of the presence of a strong cell envelope (Zhuang et al., 2003). In addition, several gram-positive high G+C strains are known to consume soluble COD (chemical oxygen demand) rapidly and store them as storage polymers to survive low nutrient conditions (Maszenan et al., 2000; Liu et al., 2001). These competitive traits can allow the gram-positive high G+C bacteria to thrive in the highly variable feast–famine situations encountered in the granulation systems, where phenol can be completely consumed within the first 30 min of each 4 h cycle (Jiang et al., 2004a).

DGGE analysis of amplified 16S rRNA gene fragments from activated sludge, granules, and isolates showed that the dominant DGGE bands

Table 9.1. Characteristics of ten strains isolated from phenol-degrading granules

Isolates	Minimum cell density (CFU g VSS^{-1})	Closest relative	Taxon. affiliation	16S rRNA gene sequence identity (%)	Number of bases analyzed
PG-01	$5.64 \pm 0.87 \times 10^{10}$	*Pandoraea apista* strain LMG 16407	β-*Proteobacteria*	98.7	1326
PG-02	$1.01 \pm 0.92 \times 10^{8}$	*Propioniferax innocua* ATCC 49929	Actinobacteria, HGC Gram positive bacteria, *Propionibacteriacaea*	93.5	1315
PG-03	$5.49 \pm 1.80 \times 10^{6}$	*Rhodococcus erythropolis* strain HV1 00/50/6670	Actinobacteria, HGC Gram positive bacteria, *Nocardioidaceae*	99.8	1433
PG-04	$3.05 \pm 1.42 \times 10^{6}$	*Propionibacterium cyclohexanicum* strain IAM 14535	Actinobacteria, HGC Gram positive bacteria, *Propionibacteriacaea*	87.7	1370
PG-05	$1.53 \pm 1.37 \times 10^{7}$	*Xenophilus azovorans* KF46FT	β-*Proteobacteria*	98.8	1437
PG-06	$2.55 \pm 1.32 \times 10^{6}$	*Acidovorax avenae* ATCC 29625	β-*Proteobacteria*	97.9	1437
PG-07	$1.93 \pm 0.72 \times 10^{5}$	*Xanthomonas axonopodis* strain s53	γ-*Proteobacteria*	98.1	1409
PG-08	$3.56 \pm 1.52 \times 10^{6}$	*Comamonas* sp. D22	β-*Proteobacteria*	97.0	1408
PG-09	$7.62 \pm 2.80 \times 10^{6}$	*Pigmentiphaga*	β-*Proteobacteria*	99.6	1432
PG-10	$4.56 \pm 1.72 \times 10^{6}$	*Hydrogenophaga palleronii* DSM 63	β-*Proteobacteria*	98.5	1483

associated with the activated sludge did not co-migrate with the dominant bands from the granules (Fig. 9.2). However, DGGE bands associated with strains PG-01, PG-02, and PG-08 co-migrated with bands from the aerobic granules, which were found to have partial sequences that were identical to the sequences of the corresponding isolates. These three strains therefore represented dominant populations of β-*Proteobacteria* and gram-positive high G+C group within the granule community.

Additional experimental results provided independent evidence to support the contention that PG-01 was a numerically important microorganism in the aerobic granules (Jiang et al., 2004b). Fluorescent *in situ* hybridization (FISH) with confocal laser scanning microscopy (CLSM) was used to elucidate the abundance and spatial distribution of strain PG-01 in the aerobic granules (Fig. 9.3). The granules consisted of a dense layer of bacterial cells, surrounding a less dense central region. This structural pattern was repeatedly observed in all sections analyzed. Most PG-01 cells were distributed in clusters in the outer layers of the granules. Direct counting of probe-hybridized cells after disruption of granules revealed that PG-01 cells were numerically abundant in the granules, accounting for $4.1 \pm 3.2\%$ of all bacterial cells. Furthermore, PG-01 had a high specific growth rate and high specific phenol degradation rate and these attributes might have contributed significantly to PG-01's dominant role in phenol degradation in the granules.

Enhanced Phenol Removal by Aerobic Granules

Aerobic granules are typically cultivated by using activated sludge as a starting inoculum. However, activated sludge might not be suitable for direct inoculation into a reactor that has a high input of chemical toxicity. One solution might be to use a better inoculum. Because the microbial community in the granules contains a high diversity of microorganisms, the granules themselves should possess enough physiological traits and a reservoir of functional responses to make them ideal candidates for use as a starting seed to rapidly produce stable granules that can efficiently degrade toxic chemicals such as phenol. In addition, the strong and compact structure of the acetate-fed granules should provide adequate protection against exposure to chemical toxicity.

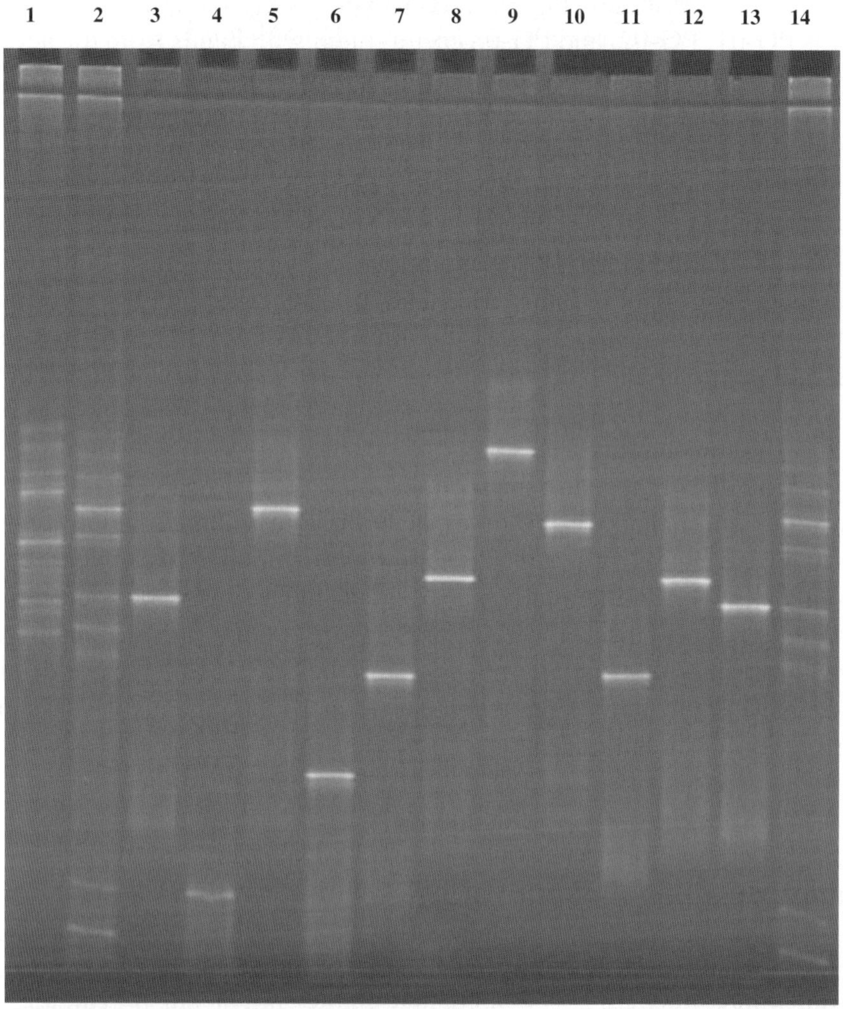

Fig. 9.2. An ethidium bromide-stained 10% polyacrylamide denaturing gradient gel (30–70%) with DGGE profiles of 16S rRNA gene fragments after PCR amplification of nucleic acids derived from acclimated activated sludge, from aerobic granules and from individual isolates. Lanes 1, activated sludge; 2, aerobic granules; 3, PG-01; 4, PG-02; 5, PG-08; 6, PG-03; 7, PG-04; 8, PG-05; 9, PG-06; 10, PG-07; 11, PG-09; 12, PG-10; 13, PG-01; 14, aerobic granules. Bands from lanes 3, 4, and 5 (strains PG-01, PG-02, and PG-08) co-migrated with bands from lane 2 (aerobic granules).

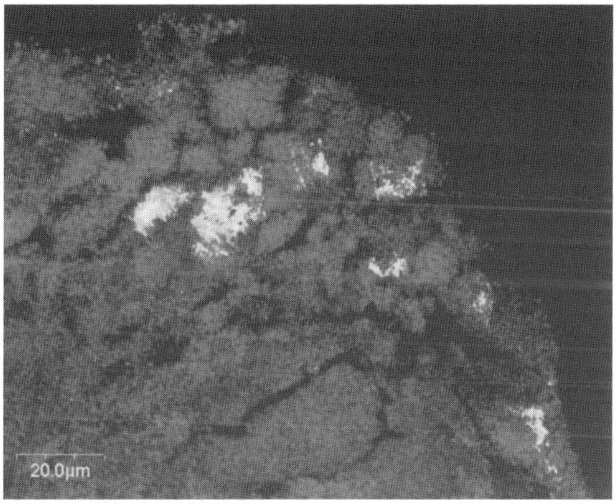

Fig. 9.3. FISH–CLSM image of outer section of the granule. Red area represents cells hybridized with an eubacterial probe and green area represents cells hybridized with a probe specific for strain PG-01 (Jiang et al., 2004b). (See Color Plate Section before the Index.)

A recent study (Tay et al., 2005) investigated the feasibility of using aerobic acetate-fed granules as a starting seed material to rapidly develop stable aerobic phenol-degrading granules. In this study, aerobic granules were first cultivated in four sequencing batch reactors with acetate as sole carbon source at a loading rate of 3.8 g l^{-1} d^{-1}. Phenol was then added to the four reactors at loading rates of 0, 0.6, 1.2, and 2.4 g l^{-1} d^{-1}, respectively. The granules acclimated quickly to the phenol loading, and stabilized only one week after phenol was introduced. The granules exhibited good settling ability with good biomass retention and good metabolic activity, as evidenced by the low SVI values, stable biomass concentrations and good removal of acetate and phenol. No significant inhibitory effects from phenol toxicity were observed at the intermediate loadings of 0.6 and 1.2 g phenol l^{-1} d^{-1}. At the highest loading of 2.4 g phenol l^{-1} d^{-1}, a sharp buildup of phenol was observed in the reactor but this quickly dissipated as the granules adapted rapidly to the high phenol concentrations. The compact structure of the acetate-fed granules likely protected the microorganisms against phenol toxicity and facilitated microbial acclimation towards faster phenol degradation rates. This concept of using granules to produce different granules can be extended to granule-based applications involving other toxic chemicals and other types of high-strength industrial wastewaters, where rapid reactor start-up and system stability are key considerations.

References

Bastos, A.E.R., Cassidy, M.B., Trevors, J.T., Lee, H., & Rossi, A. (2001). Introduction of green fluorescent protein gene into phenol-degrading Alcaligenes faecalis cells and their monitoring in phenol-contaminated soil. *Appl. Microbiol. Biotechnol.*, *56* (1–2), 255–260.

Benndorf, D., Loffhagen, N., & Babel, W. (2001). Protein synthesis patterns in Acinetobacter calcoaceticus induced by phenol and catechol show specificities of responses to chemostress. *FEMS Microbiol. Lett.*, *200* (2), 247–252.

Beun, J.J., Hendriks, A., van Loosdrecht, M.C.M., Morgenroth, E., Wilderer, P.A., & Heijnen J.J. (1999). Aerobic granulation in a sequencing batch reactor. *Water Res.*, *33* (10), 2283–2290.

Beun, J.J., van Loosdrecht, M.C.M., & Heijnen, J.J. (2002). Aerobic granulation in a sequencing batch airlift reactor. *Water Res.*, *36* (3), 702–712.

Bond, P.L., Hugenholtz, P., Keller, J., & Blackall, L.L. (1995). Bacterial community structures of phosphate-removing and non-phosphate-removing activated sludges from sequencing batch reactors. *Appl. Environ. Microbiol.*, *61* (5), 1910–1916.

Boyd, T.J., & Carlucci, A.F. (1993). Degradation rates of substituted phenols by natural-populations of marine-bacteria. *Aquat. Toxicol.*, *25* (1–2), 71–82.

Brown, V.M., Jordan, D.H.M., & Tiller, B.A. (1967). The effect of temperature on the acute toxicity of phenol in rainbow trout in hard water. *Water Res.*, *1*, 587–589.

Diaz, M.P., Boyd, K.G., Grigson, S.J.W., & Burgess, J.G. (2002). Biodegradation of crude oil across a wide range of salinities by an extremely halotolerant bacterial consortium MPD-M, immobilized onto polypropylene fibers. *Biotechnol. Bioeng.*, *79* (2), 145–153.

Fang, H.H.P., Xu, L.C., & Chan, K.Y. (2002). Effects of toxic metals and chemicals on biofilm and biocorrosion. *Water Res.*, *36* (19), 4709–4716.

Fedorak, P.M., & Hrudey, S.E. (1988). Anaerobic degradation of phenolic compounds with application to treatment of industrial waste waters. *Biotreatment Systems* (ed. Wise, D.L.), CRC Press, Boca Raton, Florida, 170–212.

Filonov, A.E., Duetz, W.A., Karpov, A.V., Gaiazov, R.R., Kosheleva, I.A., Breure, A.M., Filonova, I.F., vanAndel, J.G., & Boronin, A.M. (1997).

Competition of plasmid-hearing Pseudomonas putida strains catabolizing naphthalene via various pathways in chemostat culture. *Appl. Microbiol. Biotechnol.*, *48* (4), 493–498.

Ghisalba, O. (1983). Microbial degradation of chemical waste, an alternative to physical methods of waste disposal. *Experientia*, *39*, 1247–1257.

Harayama, S., Kok, M., & Neidle, E.L. (1992). Functional and Evolutionary Relationships among Diverse Oxygenases. *Annu. Rev. Microbiol.*, *46*, 565–601.

Heipieper, H.J., Keweloh, H., & Rehm, H.J. (1991). Influence of Phenols on Growth and Membrane-Permeability of Free and Immobilized Escherichia-Coli. *Appl. Environ. Microbiol.*, *57* (4), 1213–1217.

Heipieper, H.J., Diefenbach, R., & Keweloh, H. (1992). Conversion of Cis Unsaturated Fatty-Acids to Trans, a Possible Mechanism for the Protection of Phenol-Degrading Pseudomonas-Putida P8 from Substrate Toxicity. *Appl. Environ. Microbiol.*, *58* (6), 1847–1852.

Jiang, H.L., Tay, J.H., & Tay, S.T.L. (2002). Aggregation of immobilized activated sludge cells into aerobically grown microbial granules for the aerobic biodegradation of phenol. *Lett. Appl. Microbiol.*, *35* (5), 439–445.

Jiang, H.L., Tay, J.H., & Tay, S.T.L. (2004a). Changes in structure, activity and metabolism of aerobic granules as a microbial response to high phenol loading. *Appl. Microbiol. Biotechnol.*, *63* (5), 602–608.

Jiang, H.L., Tay, J.H., Maszenan, A.M., & Tay, S.T.L. (2004b). Bacterial diversity and function of aerobic granules engineered in a sequencing batch reactor for phenol degradation. *Appl. Environ. Microbiol.*, *70* (11), 6767–6775.

Kape, R., Parniske, M., & Werner, D. (1991). Chemotaxis and Nod Gene Activity of Bradyrhizobium-Japonicum in Response to Hydroxycinnamic Acids and Isoflavonoids. *Appl. Environ. Microbiol.*, *57* (1), 316–319.

Karlsson, A., Ejlertsson, J., & Svensson, B.H. (2000). CO2-dependent fermentation of phenol to acetate, butyrate and benzoate by an anaerobic, pasteurised culture. *Arch. Microbiol.*, *173* (5–6), 398–402.

Keweloh, H., Heipieper, H.J., & Rehm, H.J. (1989). Protection of bacteria against toxicity of phenol by immobilization in calcium alginate. *Appl. Microbiol. Biotechnol.*, *31*, 383–389.

Keweloh, H., Diefenbach, R., & Rehm, H.J. (1991). Increase of Phenol Tolerance of Escherichia-Coli by Alterations of the Fatty-Acid Composition of the Membrane-Lipids. *Arch. Microbiol.*, *157* (1), 49–53.

Kibret, M., Somitsch, W., & Robra, K.H. (2000). Characterization of a phenol degrading mixed population by enzyme assay. *Water Res.*, *34* (4), 1127–1134.

Kiesel, B., & Muller, R.H. (2002). The meta pathway as a potential energy-generating sequence and its effects on the growth rate during the utilisation of aromatics. *Acta Biotechnol.*, *22* (3–4), 221–234.

Lehman, R.M., Colwell, F.S., & Bala, G.A. (2001). Attached and unattached microbial communities in a simulated basalt aquifer under fracture- and porous-flow conditions. *Appl. Environ. Microbiol.*, *67* (6), 2799–2809.

Liu, W.T., Nielsen, A.T., Wu, J.H., Tsai, C.S., Matsuo, Y., & Molin, S. (2001). In situ identification of polyphosphate- and polyhydroxyalkanoate-accumulating traits for microbial populations in a biological phosphorus removal process. *Environ. Microbiol.*, *3* (2), 110–122.

Loh, K.C., Chung, T.S., & Ang, W.F. (2000). Immobilized-cell membrane bioreactor for high-strength phenol wastewater. *J. Environ. Eng.-ASCE*, *126* (1), 75–79.

Mace, S., & Mata-Alvarez, J. (2002). Utilization of SBR technology for wastewater treatment: An overview. *Ind. Eng. Chem. Res.*, *41* (23), 5539–5553.

Maszenan, A.M., Seviour, R.J., Patel, B.K.C., Schumann, P., Burghardt, J., Tokiwa, Y., & Stratton, H.M. (2000). Three isolates of novel polyphosphate-accumulating Gram-positive cocci, obtained from activated sludge, belong to a new genus, *Tetrasphaera* gen. nov., and description of two new species, *Tetrasphaera japonica* sp, nov and *Tetrasphaera australiensis* sp nov. *Int. J. Syst. Evol. Microbiol.*, *50*, 593–603.

Morgenroth, E., Sherden, T., van Loosdrecht, M.C.M., Heijnen, J.J., & Wilderer, P.A. (1997). Aerobic granular sludge in a sequencing batch reactor. *Water Res.*, *31* (12), 3191–3194.

Moslemy, P., Neufeld, R.J., & Guiot, S.R. (2002). Biodegradation of gasoline by gellan gum-encapsulated bacterial cells. *Biotechnol. Bioeng.*, *80* (2), 175–184.

Moy, B.Y.P., Tay, J.H., Toh, S.K., Liu, Y., & Tay, S.T.L. (2002). High organic loading influences the physical characteristics of aerobic sludge granules. *Lett. Appl. Microbiol.*, *34* (6), 407–412.

Muller, R.H., & Babel, W. (1996). Growth rate dependent expression of phenol assimilation pathways in Alcaligenes eutrophus JMP 134 – The influence of formate as an auxiliary energy source on phenol conversion characteristics. *Appl. Microbiol. Biotechnol.*, *46* (2), 156–162.

Ng, L.C., Poh, C.L., & Shingler, V. (1995). Aromatic effector activation of the Ntrc-like transcriptional regulator Phhr limits the catabolic potential of the (methyl)phenol degradative pathway it controls. *J. Bacteriol.*, *177* (6), 1485–1490.

Nordlund, I., Powlowski, J., Hagstrom, A., & Shingler, V. (1990). Complete nucleotide and polypeptide analysis of multi-component phenol hydroxylase from Pseudomonas sp. strain CF600. *J. Bacteriol.*, *172*, 6826–6833.

Okuyama, H., Okajima, N., Sasaki, S., Higashi, S., & Murata, N. (1991). The Cis Trans Isomerization of the Double-Bond of a Fatty-Acid as a Strategy for Adaptation to Changes in Ambient-Temperature in the Psychrophilic Bacterium, Vibrio Sp Strain Abe-1. *Biochim. Biophys. Acta*, *1084* (1), 13–20.

Prieto, M.B., Hidalgo, A., Rodriguez-Fernandez, C., Serra, J.L., & Llama, M.J. (2002). Biodegradation of phenol in synthetic and industrial wastewater by Rhodococcus erythropolis UPV-1 immobilized in an air-stirred reactor with clarifier. *Appl. Microbiol. Biotechnol.*, *58* (6), 853–859.

Rittmann, B.E., & McCarty, P.L. (2001). *Environmental Biotechnology: Principles and Applications*, New York: McGraw-Hill.

Smith, E.A., & Macfarlane, G.T. (1997). Formation of phenolic and indolic compounds by anaerobic bacteria in the human large intestine. *Microb. Ecol.*, *33* (3), 180–188.

Snaidr, J., Amann, R., Huber, I., Ludwig, W., & Schleifer, K.H. (1997). Phylogenetic analysis and in situ identification of bacteria in activated sludge. *Appl. Environ. Microbiol.*, *63* (7), 2884–2896.

Tay, J.H., Liu, Q.S., & Liu, Y. (2001). The effects of shear force on the formation, structure and metabolism of aerobic granules. *Appl. Microbiol. Biotechnol.*, *57* (1–2), 227–233.

Tay, S.T.L., Ivanov, V., Yi, S., Zhuang, W.Q., & Tay, J.H. (2002). Presence of anaerobic bacteroides in aerobically grown microbial granules. *Microb. Ecol.*, *44* (3), 278–285.

Tay, J.H., Tay, S.T.L., Ivanov, V., Pan, S., Jiang, H.L., & Liu, Q.S. (2003). Biomass and porosity profiles in microbial granules used for aerobic wastewater treatment. *Lett. Appl. Microbiol.*, *36* (5), 297–301.

Tay, J.H., Jiang, H.L., & Tay, S.T.L. (2004) High-rate biodegradation of phenol by aerobically grown microbial granules. *J. Environ. Eng.-ASCE*, *130* (12), 1415–1423.

Tay, S.T.L., Moy, B.Y.P., Jiang, H.L., & Tay, J.H. (2005). Rapid cultivation of stable aerobic phenol-degrading granules using acetate-fed granules as microbial seed. *J. Biotechnol.*, *115* (4), 387–395.

Tresse, O., Lorrain, M.J., & Rho, D. (2002). Population dynamics of free-floating and attached bacteria in a styrene-degrading biotrickling filter analyzed by denaturing gradient gel electrophoresis. *Appl. Microbiol. Biotechnol.*, *59* (4–5), 585–590.

van Schie, P.M., & Young, L.Y. (2000). Biodegradation of phenol: mechanisms and applications. *Bioremediation J.*, *4* (1), 1–18.

Villaverde, S., & Fernandez-Polanco, F. (1999). Spatial distribution of respiratory activity in Pseudomonas putida 54G biofilms degrading volatile organic compounds (VOC). *Appl. Microbiol. Biotechnol.*, *51* (3), 382–387.

Watanabe, K., Hino, S., & Takahashi, N. (1996). Responses of activated sludge to an increase in phenol loading. *J. Ferment. Bioeng.*, *82* (5), 522–524.

Watanabe, K., Teramoto, M., & Harayama, S. (1999). An outbreak of nonflocculating catabolic populations caused the breakdown of a phenol-digesting activated-sludge process. *Appl. Environ. Microbiol.*, *65* (7), 2813–2819.

Watanabe, K., Teramoto, M., Futamata, H., & Harayama, S. (1998). Molecular detection, isolation, and physiological characterization of functionally dominant phenol-degrading bacteria in activated sludge. *Appl. Environ. Microbiol.*, *64* (11), 4396–4402.

Whiteley, A.S., Wiles, S., Lilley, A.K., Philp, J., & Bailey, M.J. (2001). Ecological and physiological analyses of Pseudomonad species within a phenol remediation system. *J. Microbiol. Methods*, *44* (1), 79–88.

Wilderer, P.A., Irvine, R.L., & Goronszy, M.C. (2001). *Sequencing Batch Reactor Technology*, IWA Publishing, London.

Yap, L.F., Lee, Y.K., & Poh, C.L. (1999). Mechanism for phenol tolerance in phenol-degrading Comamonas testosteroni strain. *Appl. Microbiol. Biotechnol.*, *51* (6), 833–840.

Yoong, E.T., Lant, P.A., & Greenfield, P.F. (2000). In situ respirometry in an SBR treating wastewater with high phenol concentrations. *Water Res.*, *34* (1), 239–245.

Zhuang, W.Q., Tay, J.H., Maszenan, A.M., Krumholz, L.R., & Tay, S.T.L. (2003). Importance of Gram-positive naphthalene-degrading bacteria in oil-contaminated tropical marine sediments. *Lett. Appl. Microbiol.*, *36* (4) 251–257.

Chapter 10

Seeds for Aerobic Microbial Granules

Volodymyr Ivanov and Stephen Tiong-Lee Tay

Advantages of Microbial Granulation

Microbial granulation is a process exploited in biological wastewater treatment whereby bacteria are organized into highly structured suspended granules that are capable of removing biodegradable organic matter, nitrogen, and phosphorus. Parts of microbial granule, probably, have coordinated physiological functions, i.e. cell growth, metabolism, interactions, biosynthesis, transport, consumption, and storage of nutrients.

Microbial granulation differentiated from flocculation and formation of microbial flocs by the following definition: granules making up granular activated sludge are aggregates of microbial origin (no carrier material is intentionally involved or added), which do not coagulate under reduced hydrodynamic shear, and which settle significantly faster than activated sludge flocs (de Kreuk et al., 2005). Microbial granules are usually spheres with diameter from 0.5 to 4 mm. Microbial flocs formed in conventional wastewater biological treatment due to the recycling from secondary settling tank are loose aggregates with undefined shape and size from 0.05 to 0.2 mm. Microbial granules are formed under aerobic conditions with such selection factors as settling time from 2 to 10 min (Tay et al., 2001c; Qin et al., 2004) and high aeration rate ensuring superficial upflow air velocity above 1.2 cm/s in a column sequencing batch reactor (SBR) (Tay et al., 2001b; Liu and Tay, 2002, 2004). The primary aim of the formation of strong microbial granules and their application in industrial and municipal wastewater treatment is to avoid the construction of secondary settling tanks or to diminish their size. Formed granules have not only settling velocity higher than $10\,m\,h^{-1}$ (Beun et al., 1999) but showed several other advantages over microbial flocs of conventional activated sludge, including reduced biomass yield (Tay et al., 2003b) and higher resistance to toxic compounds due to its compact structure (Glancer et al., 1994; Jiang et al., 2002; Bergsma-Vlami et al., 2005; Tay et al., 2005). Therefore, activated granular sludge systems are developing for the treatment of industrial wastewater and for application in the places where land is a premium. Aerobic granular sludge has been successfully used to treat real industrial wastewater like dairy effluents (Arrojo et al., 2004; Schwarzenbeck et al., 2005).

Disadvantages of Microbial Granulation

However, one disadvantage of aerobic granulation is a long start-up period of granule formation from the flocs of activated sludge. The formation of aerobic granules is very crucial for their applicability in wastewater treatment, while this process takes several weeks to start-up aerobic granular system from conventional activated sludge (Peng et al., 1999; Beun et al., 2000; Tay et al., 2001c; Moy et al., 2002). Another potential disadvantage is the risk of accumulation of pathogenic microorganisms in the granule because of two reasons:

1. Cells are aggregated mainly due to hydrophobic interactions and there may be accumulation of strains with high cell hydrophobicity in granule;
2. Bacterial strains with high cell surface hydrophobicity are often pathogenic ones.

Principles of Facilitated Granule Formation

A priori, a lot of methods can be applied to facilitate bacterial cells aggregation and formation of microbial granules:

1. addition of flocculants;
2. change of pH;

3. addition of suspended carrier for the formation of granule with particle inside;
4. addition of previously formed granules;
5. addition of disrupted, previously formed granules;
6. addition of enrichment culture of fast aggregating cells;
7. addition of pure culture of fast aggregating cells;
8. optimal aeration;
9. optimal mechanical effects facilitating cell aggregation (mechanical granulation).

Cell Aggregation by Application of Reagents and Adsorbents

It is well-known that aggregation of bacterial cells and formation of flocs and even more dense aggregates can be enhanced by addition of such flocculants as calcium, aluminum, and iron ions or organic flocculants. These reagents form salt bridges between cell surfaces, adsorb or connect cells due to electrostatic interactions between charges of inorganic or organic flocs and cell surface (Calleja, 1984). However, cell aggregates will be of irregular shape with different sizes and settling velocities. It can also be too expensive a method of cell aggregation for large-scale wastewater treatment. Discharge of flocculant-containing effluent cannot be safe for environment. Decrease of pH to 4–5 can neutralize net charge of cell surface due to neutralization of carboxylic groups. It is facilitating cell aggregation due to decrease or electrostatic repulsion and increase of hydrophobic interactions. However, it can be applied only to enhance concentration of microbial biomass but not the process of microbial cultivation because optima of pH for aggregation and growth are different.

Particle-based biofilm reactors provide the potential to develop compact and high-rate processes. In these reactors, a large biomass content can be maintained (up to $30\,g\ L^{-1}$), and the large specific surface area (up to $3000\,m^{-1}$) ensures that the conversions are not strongly limited by the biofilm liquid mass-transfer rate. Engineered design and control of particle-based biofilm reactors are established, and reliable correlations exist for the estimation of the design parameters. As a result, a new generation of high-load, efficient biofilm reactors are operating throughout the world with several full-scale applications for industrial and municipal wastewater treatment (Nicolella et al., 2000).

Granules as Seeds for Granulation

Addition of previously grown microbial biomass as starter culture is a common approach in wastewater treatment plants to start up the conventional activated sludge and anaerobic digestion processes after technological accidents or process failure. Therefore, similar approach can be used to start up the process with granulated microbial biomass. There are some known commercial application of the granules as the seeds to upgrade or initiate wastewater treatment, for example, product ARGUS® (EcoEngineering Ltd., Nova Vas, Croatia) consisting of granules used as the seeds for biological treatment of wastewater from chemical, pharmaceutical, and food industries, as well as oil refineries, landfills, pig, and poultry farms in cases that existing treatment plant is not working properly and should be upgraded and new plant for treatment of complex and toxic chemicals such as phenols, antibiotics, lignosulfonates, naphthalene, and high concentration of nitrogen compounds. The granules have high settling velocity and contain various strains of microorganisms, which are able to accept toxic shocks and perform different physiological functions useful for wastewater treatment. There are no data on the stability of inoculated granules in the conventional wastewater treatment system. Though, it would be reasonable to expect that the inoculated granules will be replaced by more dispersed aggregates after some time of cultivation if there will be no special selection and retention of the granules during wastewater treatment.

Life Cycle of the Granule and Determination of Retention Time for the Granules in SBR

Probably, addition of disrupted, previously formed granules can also be a method for facilitation of granulation. It can be useful in case the life cycle of the granule is short and new granules are produced not from the existing granules but from the cells or particles of disrupted granules.

Attachment and removal of granule biomass was evaluated from the curves showing the content of the fluorescent lipophilic tracer in biomass of granules (F_g) and flocs (F_s) for period of cultivation between 2 and 6 days after labeling (Fig. 10.1). Concentration of granular biomass (VSS) during that period was stable, at $6.5 \pm 0.2\,g\,L^{-1}$. Concentration of floc biomass (VSS) was $0.15 \pm 0.02\,g\,L^{-1}$. The hydraulic residence time was 0.33 d, which corresponded to a daily exchange of three reactor volumes. Therefore, the ratio of produced granular biomass to produced floccular biomass was 14.5. This ratio was close to 18.3, the initial ratio of granular labeled biomass to the flocculent labeled biomass after 4 h of labeling (one growth cycle in SBR is 4 h). The granules were retained in the SBR while the flocs were washed out with the effluent. Therefore, stable concentration of granular biomass can be due to the balanced attachment and detachment of the flocs to granules. Content of lipophilic tracer in granular biomass was stable for 6 days of study (Fig. 10.1) and was thought to be because of balanced attachment and detachment of the flocs to granules or balanced growth and destruction of the granules. It cannot be the result of negligible degradation of granules because the labeled biomass permanently released as the labeled flocs. The tracer content could decrease if the rate of granule growth is higher than the rate of granule degradation.

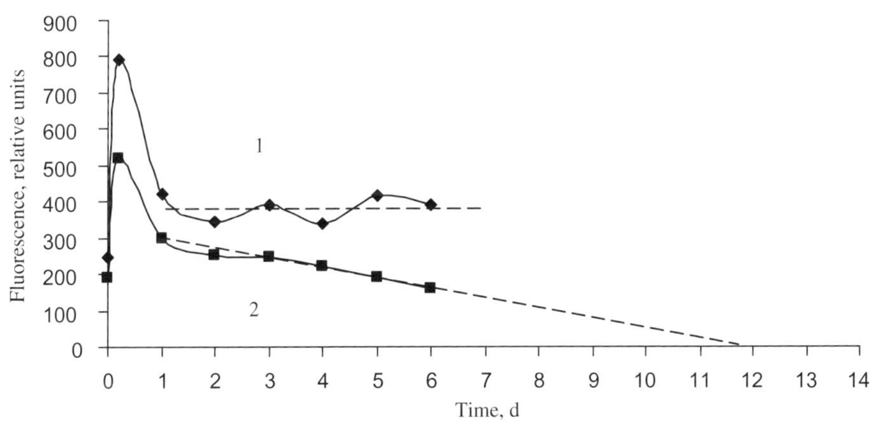

Fig. 10.1. Changes in fluorescence of the lypophilic label in the column sequencing batch reactor (curve 1, label of fast settling granules) and in the effluent (curve 2, fluorescence of sludge $\times$ 10). Dashed lines show the trend of fluorescence of the granules and the flocs.

The duration of the life cycle of the granule was evaluated by determining the intersection of the extrapolated line of F_s with the time axis assuming that this is the time (t_g) taken for all labeled material to be detached from the granules and removed from the SBR. The value of t_g was 11.8 days for detachment of 100% of labeled material (or degradation of 100% of labeled granules). This means that labeled matter could disappear from the granules completely after approximately 12 days, i.e. that lifetime for the studied granules or for the labeled matter of these granules was 12 days. A similar duration is required to transform flocs into granules. If the lifetime of the granules (not just lifetime of their labeled matter) is 12 days, the retention time for the granules must be probably less than their lifetime to avoid degradation of the granules. Therefore, it would be better for stable wastewater treatment process to use particles from disrupted granules as the seeds to ensure their further growth to the matured and stable granules, however.

Selection of Microbial Seeds from Granules

The efficiency of biological wastewater treatment depends on the growth of metabolically capable microorganisms and efficient separation of those organisms from the treated effluent. Bacterial cells used in conventional wastewater treatment aggregate and form flocs. To separate these flocs in conventional activated sludge system, a big secondary sedimentation tank is required because of relatively slow settling velocity of sludge flocs. In contrast, microbial granules settle significantly faster. The average settling velocity of microbial granules with a diameter of 3.2 mm was 0.97 cm s^{-1} (Etterer and Wilderer, 2001). This good settleability of the granules makes settling tanks superfluous (de Bruin, 2004; de Kreuk and van Loosdrecht, 2004). The benefits expected from aerobic granulation are compact treatment plants and simple reactor design (de Kreuk et al., 2005).

The purpose of this research was to select aggregate-associated bacterial cultures from microbial granules and to examine their ability to accelerate formation of granules during wastewater treatment. One way to achieve this goal was to isolate small aggregates, to disperse them, and then to study reaggregation. However, it was found by Snidaro et al. (1997) that it would be unlikely to disrupt totally the microcolonies of activated sludge flocs without significant cell lyses because cells are tightly bound together by a gel matrix. These microcolonies had a medium diameter of 13 μm and were linked by polymers (Li and Ganczarczyk, 1990; Jorand et al., 1995; Snidaro et al., 1997). Therefore, the idea of our experiments was to select self-formed microcolonies after destruction of granules, to separate microcolonies/microaggregates by fast settling, and then to grow them in fresh medium. By repeating this selection procedure, aggregates-forming microbial culture was enriched, and microbial strains with high aggregation ability have been isolated. Cell aggregation in enrichment culture appeared during stationary phase of batch cultivation. It was suggested, the depletion of nutrients could stimulate cell aggregation. It is known that under starvation, bacterial cell surface become more hydrophobic and it might facilitate cell aggregation (Bossier and Verstraete, 1996).

Microbial granulation is an autoselection process, *a priori* causing accumulation of cells with high aggregation ability in formed granules. Therefore, these cells could be selected, isolated, selected, and used to start up a facilitated granulation process.

Microbial cells with high cell surface hydrophobicity and high settleability were selected from the disrupted granules. The granules were taken from a reactor, disrupted in a beater for 2 min, and then the disrupted granules were filtered through a 25-μm pore membrane. Two kinds of microaggregates produced were studied after 30 min of settling. One type of microaggregates, with high hydrophobicity, was accumulated in the biofilm attached to the water–air interphase. Another type of microaggregates, with high settling velocity, settled down and accumulated on the bottom of the tube. The size distributions of these microaggregates were different (Fig. 10.2). Microaggregates with high hydrophobicity had narrow size distribution with mean diameter of particles 3 μm, while diameter of particles without any selection (cells from the bulk of suspension) was 2 μm. Particles with high settleability had wider size distribution with mean diameter of particles 6 μm. Fast formation of two types of cell aggregates from microbial granules was used for selection of microbial seeds facilitating formation of microbial granules.

Use of Enrichment Culture for Facilitated Granule Formation

By the analogy with other wastewater treatment systems, formation of the granules can be enhanced by selected microbial cultures (Beun et al., 1999). For example, Limbergen et al. (1998) proposed that selection and

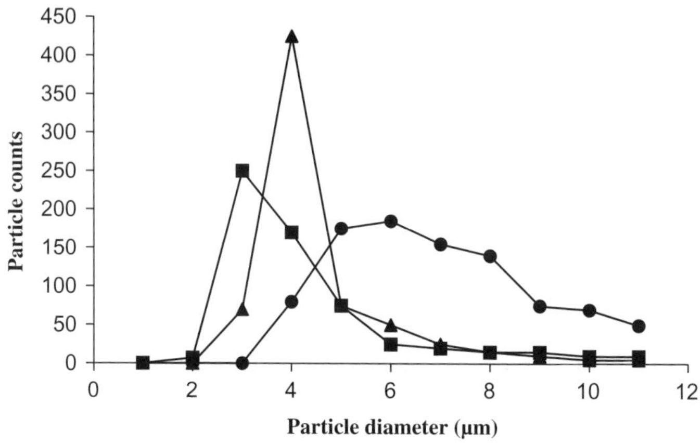

Fig. 10.2. The size distribution of three kinds of different cells from disrupted granules. Selected cells with high settleability (•); high hydrophobicity (▲); cells before selection (■) are shown.

application of floc-forming bacteria are important for good flocculation in activated sludge system. When the inoculated microorganisms are incorporated into activated sludge, they can stay in this aggregate for a longer time, thus helps to form floc and maintain the degradative capacity of the flocs. Addition of selected strains or enrichment culture with specific function was also helpful in wastewater treatment (Dabert et al., 2005).

Ivanov et al. (2005) applied enrichment culture with increased cell surface hydrophobicity for faster formation of the granules. Mechanically stronger granules, which were suitable for the reactors with mechanical stirring, were formed for several days after the start of the cultivation. The strains of aerobic bacteria with aggregation index (AI) higher than 60–80% or with cell hydrophobicity, measured by hydrocarbon adherence test, higher than 80%, were selected and isolated from the microbial granules using the repeated cycles of adhesion, settling, and cultivation. There were sporogenic gram-positive rods, gram-negative rods, and gliding bacteria.

The duration of the lifetime of the granules is close to the time of granule formation from activated sludge which is usually from 8 to 14 days (Fig. 10.3). This duration can be reduced if selected microbial cultures with high self-aggregation ability are added to the SBR. The granules with mean diameter 1 mm were formed after 2 days when cells with high cell surface hydrophobicity were used as inoculum (seeds) for granulation

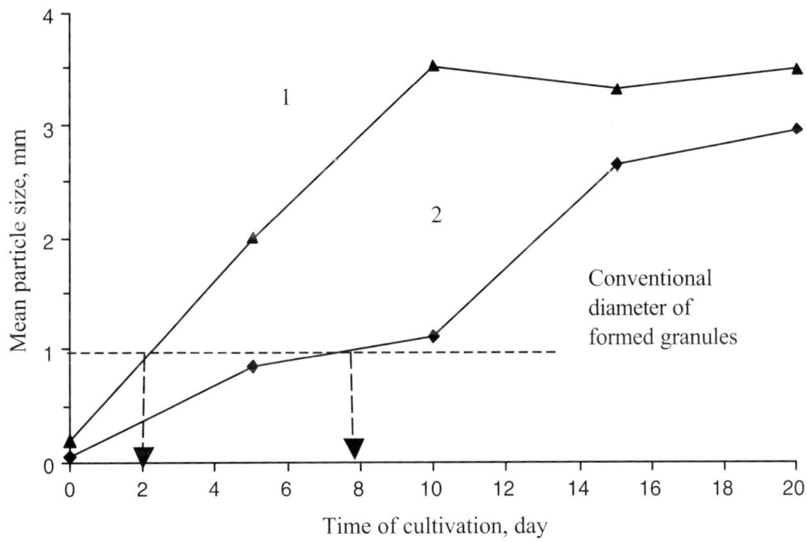

Fig. 10.3. Formation of the granules from selected cells with high cell surface hydrophobicity (1) and from the flocs of activated sludge (2).

(Fig. 10.3). Formation of the granules with mean diameter 1 mm from the flocs of activated sludge required 8 days.

Selection of Pure Cultures for Facilitated Granule Formation

The production of compact aerobic granules is favored by a short sludge settling time, which selects for bacterial aggregates with a high settling velocity (Beun et al., 1999, 2000). The formation of a stable granular structure has also been positively correlated with the strength of the hydrodynamic turbulence caused by the upflow aeration in a sequential batch reactor (Tay et al., 2001a).

The enrichment culture, with the ability of accelerated granulation, can be obtained by repeating 10 min settling and batch cultivation of fast settling microbial aggregates isolated from the aerobic granular sludge. Aggregation index (AI) of cells in the enrichment culture increased from 4 to 34% after 13 cycles of selection, and is transferred to a liquid medium (Fig. 10.4).

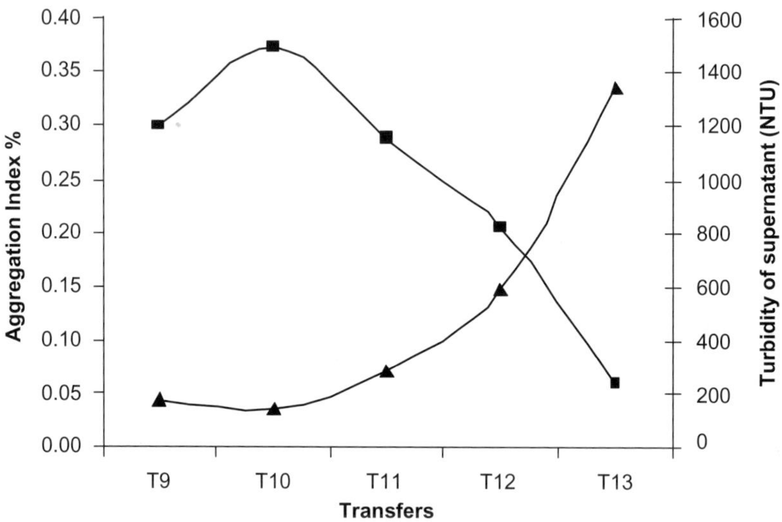

Fig. 10.4. Aggregation index and supernatant turbidity change during enrichment culture transfers. (▲): Aggregation Index; (■): Turbidity of supernatant after 10 min settling.

Enrichment culture, produced after 13 cycles of selection and transfers, was used because there were no significant changes of cell aggregation ability after 13 cycles (Fig. 10.4). During batch cultivation of enrichment culture after 13 transfers, cell aggregation increased in stationary phase (Fig. 10.5).

Both the increased AI and decreased supernatant turbidity after settling of biomass indicated the good aggregation ability of enrichment culture.

Isolation of Pure Cultures with High Self-aggregation Ability

Eleven pure cultures, distinguished by size, color, and shape of colonies, were isolated from the enrichment culture by plate-spreading technique. Two strains, B and F, with highest aggregation ability were selected from these 11 strains. After that, the sequences of 16S rRNA of these two selected strains were used for identification of species. Microbial strains

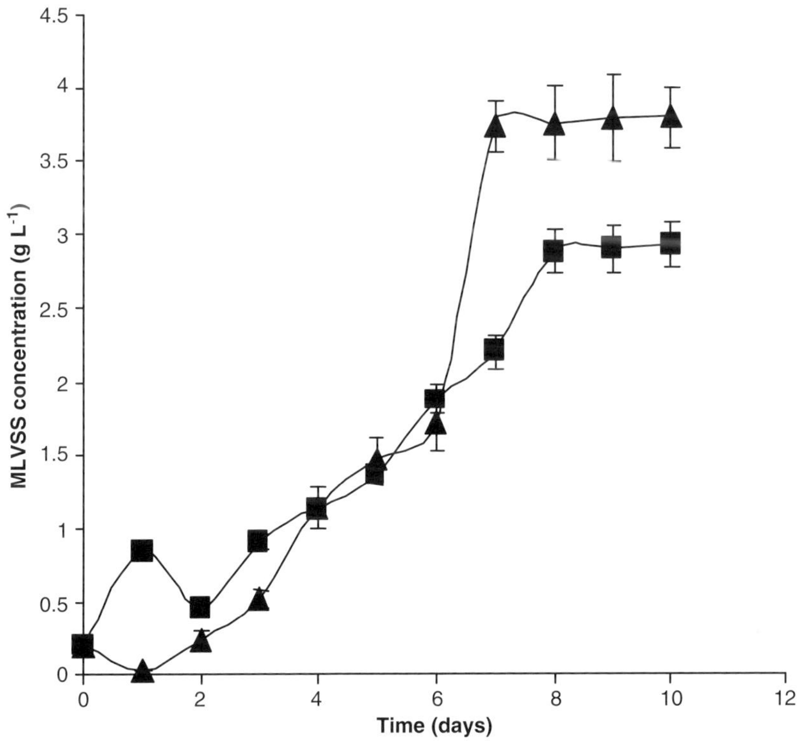

Fig. 10.5. Dynamics of aggregation index and cell biomass during batch growth of enrichment culture. (▲): Aggregation Index; (■): Turbidity of microbial suspension.

B and F had AI of 65 and 51%, respectively. Strains B and F can also be coaggregated with each other with coaggregation index of 58%. Cells of both strains were gram-negative aerobic rods.

Formation of Granules

The maximum of biomass (MLVSS) was 3.8 g L^{-1} in control and 2.9 g L^{-1} in experiment. The SVI values in control (reactor R1) were always higher than in experiment (reactor R2). After three days of cultivation, SVI in experiment (reactor R2) was 80 mL g^{-1}, which is close to typical SVI of matured granules (Fig. 10.6).

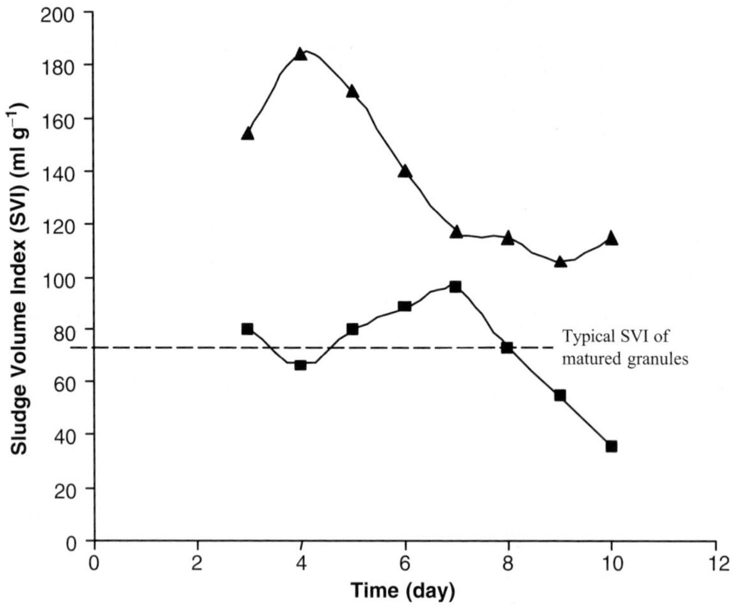

Fig. 10.6. SVI changes during cultivation. (▲): control reactor (R1); (■): experimental reactor (R2).

At the same time, SVI in control was 150 mL g^{-1}. Lowest SVI in experiment was 36 mL g^{-1} and lowest SVI in control was 110 mL g^{-1}, which is almost three times higher than the minimum for the experiment.

Aggregates in experiment (reactor R2) were mainly the granules (Fig. 10.7). The aggregates in control were the mixture of flocs, granules of regular structure, fluffy granules, bulking sludge, and filamentous bacteria (Fig. 10.7). There was no statically significant difference in the particle sizes in the two bioreactors. The presence of filamentous bacteria in control could be the major reason for the high SVI value.

Granules started to form in the experiment after three days of cultivation, while only microbial flocs can be seen at that time in control. It is showed that addition of strains B and F with higher aggregation abilities than that of activate sludge, reduced the duration of granulation process from several weeks to 8 days. This evaluation is based on maximum of biomass accumulation and value of SVI, which was lower than 70 mL g^{-1} of the typical value of matured granules (Fig. 10.6). F/M ratio in the experimental and control reactor was 1.46 g COD g VSS^{-1} day^{-1},

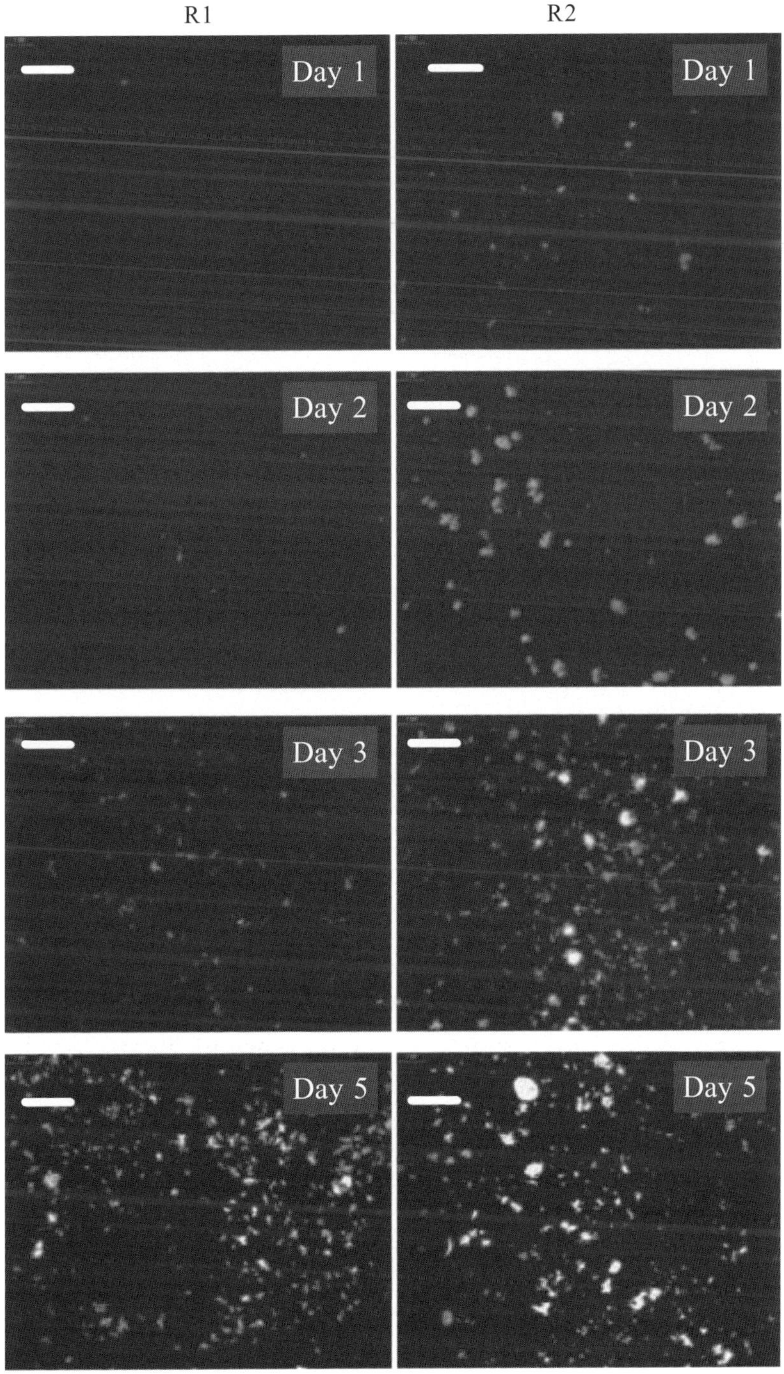

Fig. 10.7. Morphological changes during the cultivation in control (reactor R1) and experiment (reactor R2).

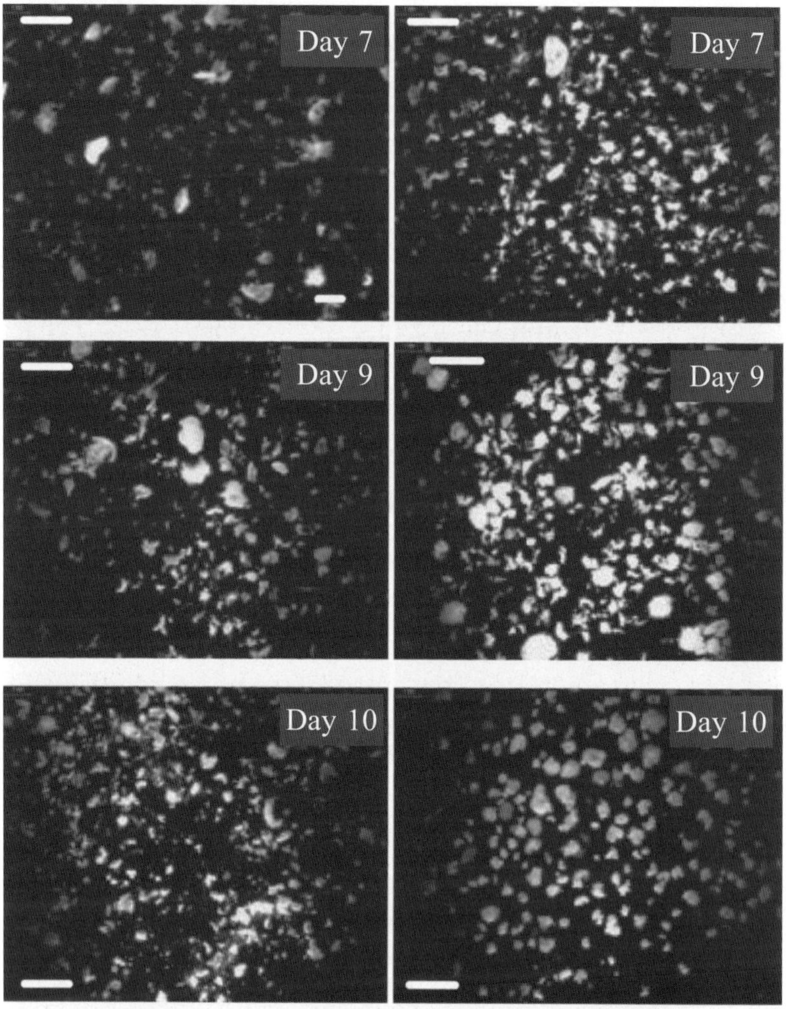

Scale bar (▬) is 1 mm.

Fig. 10.7. — Cont'd.

and 1.12 g COD g VSS^{-1} day^{-1}, respectively. Mean cell residence times were 15 and 4.7 days for experimental and control reactors, respectively. The COD removal efficiencies for both control and experiment reactors were stable at 95% after 8 days of cultivation (Fig. 10.6). The granules were formed and dominated a bioreactor after 8 days of cultivation in an experimental reactor and the COD removal efficiency was stable after 8 days of cultivation.

Microscopy and Microbiology of the Granules

The granules in experiment had compact structure, with bacterial cells in shape of rods and cocci on surface (Fig. 10.8). Small amount of filamentous bacteria was found on surface of granules. The rods and cocci were connected together by slime matrix (Fig. 10.8).

Phylogenetic Identification and Evaluation of Biosafety of Selected Strains

Full 16S rRNA gene sequences were obtained for microbial strains B and F, respectively. The sequences of strain B was 99.4% identical to *Klebsiella pneumoniae* [Gene bank access number is AY292866.1], and strain F was 99.9% identical to *Pseudomonas veronii* [Gene bank access number is AY512619.1]. The sequences were of 1338 and 1408 bases in length, respectively. Both isolates belong to subclass of γ-Proteobacteria.

A disadvantage of wastewater treatment with microbial granules in comparison with the conventional activated sludge system is the long start-up period. To accelerate the granulation start-up and to prevent the growth of filamentous bacteria, high COD loading, 8.5 g COD L^{-1} day^{-1}, was used in this study because there was no negative effect on granulation when COD increased even up to 15 kg m^{-3} day^{-1}(Moy et al., 2002). Another approach was selection of cells with high aggregation ability. Cells of selected strains *Klebsiella pneumoniae* strains B and *Pseudomonas veronii* F could either form aggregates by themselves or coaggregated with other microbial strains of activated sludge. Such ability of these strains could be the main reasons why microbial granules can be formed faster in experiment with addition of these strains to activated sludge than in control with activate sludge as inoculum.

Cultivation of cells from the dispersed granules on solid medium showed that $12 \pm 3\%$ of colonies from the granules formed for 10 days of cultivation in experimental reactor (R2) were colonies of *Klebsiella*

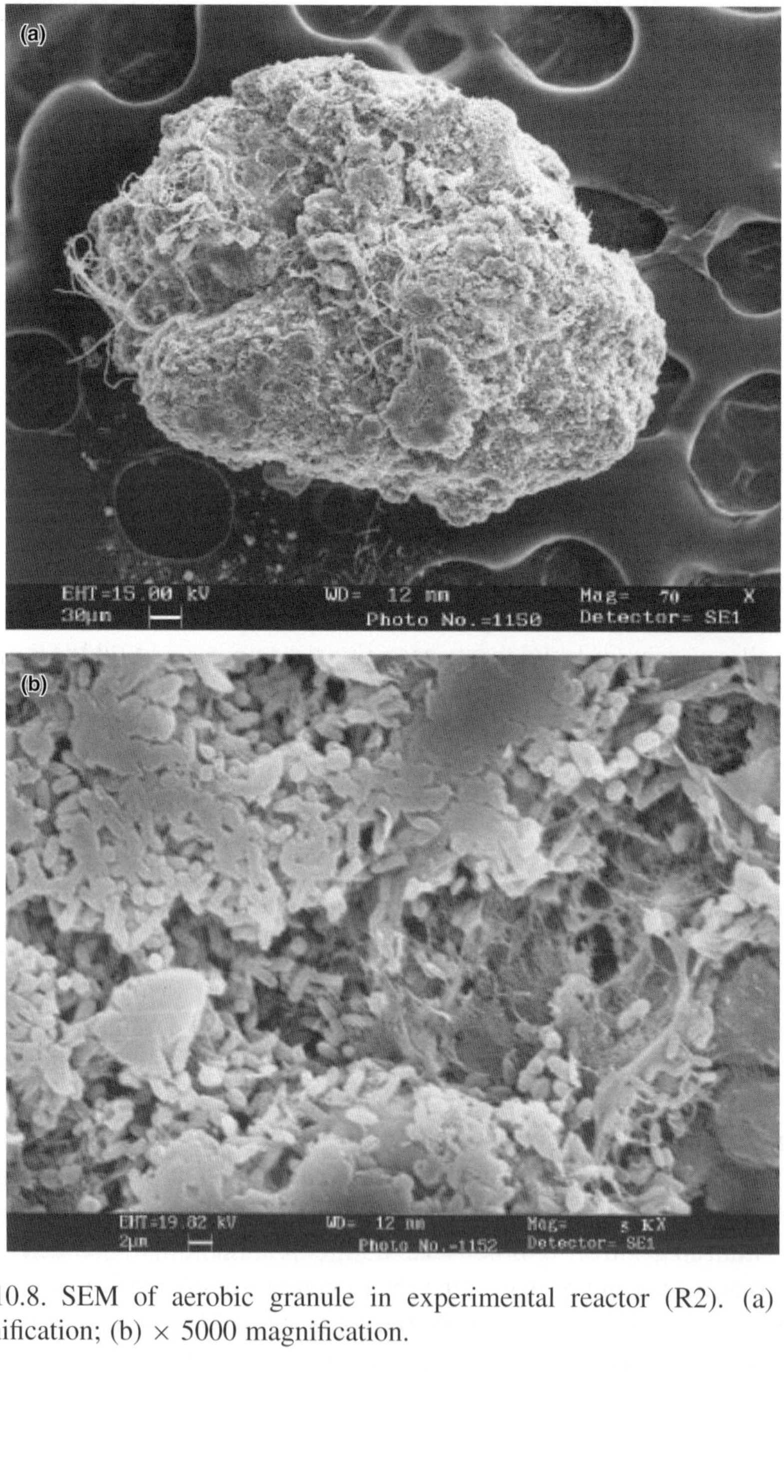

Fig. 10.8. SEM of aerobic granule in experimental reactor (R2). (a) × 70 magnification; (b) × 5000 magnification.

pneumonia strain B and 40 ± 6% colonies were colonies of *Pseudomonas veronii* strain F. The concentrations of *Klebsiella pneumoniae* and *Pseudomonas veronii* were both 36 mg L^{-1} at the beginning of cultivation. After 10 days of cultivation, these concentrations increased to 350 and 1400 mg L^{-1}, respectively. It indicated that the added strains B and F were dominant cultures in formed aerobic granules.

Bioaugmentation of activated sludge systems with specialized bacterial strains (microbial seeds), has been practiced since the 1960s. Their application in wastewater treatment was originally the efforts to solve operational problems such as shock loading in treatment plant (Limbergen et al., 1998). Bioaugmentation could be a powerful tool and cost-effective method to improve several aspects in the wastewater treatment process such as improved flocculation and degradation of recalcitrant compounds (Limbergen et al., 1998). A stable enhanced biological phosphorus removal (EBPR) in a bioreactor was installed within 15 days using bioaugmentation of sludge by the phosphorus-accumulating organism (Dabert et al., 2005). In our experiments, environmental conditions and process parameters in both control and experimental reactors were the same and only the difference between the reactors was an addition of two selected strains with high aggregation ability into experimental reactor. Experimental data demonstrated that this addition significantly reduced the time of granule formation and facilitated formation of dense granule with low SVI.

Both isolates, *Klebsiella pneumoni* strain B and *Pseudomonas veronii* strain F, belong to γ-Proteobacteria, which in agreement with previous studies that Proteobacteria constitute a largest fraction of the microbial granules (Jiang et al., 2004b). The authors isolated seven strains and one from them, *Comamonas* sp. D22, exhibited strong flocculation activity and could form auto-aggregates with high extra-polysaccharide content, which might play an important role in the formation and maintenance of the phenol-degrading aerobic granules (Jiang et al., 2004a,b).

However, the species *K. pneumoniae* is a urinary tract pathogen and could be considered as opportunistic human pathogen. The release of the strain of this species into environmental engineering system might potentially cause health problems for human and animals. Therefore, *K. pneumoniae* strain B was considered as not suitable for environmental engineering application because of biosafety issue. Strain of *Pseudomonas veronii* might be considered as suitable for the treatment of wastewater because until now there were no published data on its pathogenicity. Therefore, *Pseudomonas veronii* strain F was selected for further large-scale trials as a starter culture for facilitated formation of microbial granules in aerobic wastewater treatment.

However, survival and stable activity of introduced strain in the system is a common problem in bioaugmentation applications. Therefore, selected strain of *Pseudomonas veronii* cannot be considered as universal strain for all cases, where microbial granulation is required. The feasibility of microbial granulation start-up with this strain must be tested for every specific process. Another problem of bioaugmentation of microbial granules with selected strain is that not only duration of granulation, but also the specific activities of granules such as nitrification, spectra of degradation of natural organic compounds and xenobiotics, accumulation of phosphate, cell survivability, and other properties are important in wastewater treatment. Therefore, enhancement of specific activities of microbial granules by incorporation of other microbial strains into the granules must also be studied.

Bacterial cultures of *Klebsiella pneumoniae* strain B and *Pseudomonas veronii* strain F, with self-aggregation index of 65 and 51%, respectively, and coaggregation index of 58%, were isolated from enrichment culture. The mixture of these strains with activated sludge was used as inoculate in an experimental sequencing batch reactor to start-up aerobic granulation process. Aerobic granules with mean diameter of $446 \pm 76\,\mu m$ have been formed in experiment after 8 days of cultivation but the microbial granules were absent in control. Considering biosafety issues, *Klebsiella pneumoniae* strain B will be excluded from further studies, but *Pseudomonas veronii* strain F was selected for larger scale testing. Time of granule formation from the flocs of activated sludge was from 8 to 14 days but can be reduced to 2 days if selected bacterial strains with enhanced self-aggregation ability will be used instead of activated sludge.

Diversity of Granule versus Fast Granulation

Aerobically grown microbial granules have diverse microbial community, complex spatial structure, coordinated physiological functions, and specific temporal changes (Tay et al., 2003a,b; Ivanov et al., 2004, 2005). Using confocal laser scanning microscopy (CLSM) and fluorescence *in situ* hybridization with oligonucleotide probes, it was shown that microbial granules were composed of a variety of biological layers arranged as a sequence of obligate aerobic microorganisms, facultative anaerobic, obligate anaerobic bacteria, and finally a core of dead and lysed cells (Tay et al., 2002a,b; Ivanov et al., 2004, 2005). Granules also contain protozoa on their surface (Ivanov et al., 2004). Due to the diversity of granules and their structures, microbial granules can be used as bioagent to treat wastewater or to recover wastewater treatment system after sludge bulking or physiological shocks.

Selection of Granules with Nitrifying Activity

Figure 10.9 shows the time course of process by SBR operation. After 17 days of start-up, no nitrification occurred. At day 17, the reactor was inoculated with a little of nitrifying sludge. From day 18, weak nitrification happened. After another 3 weeks, the granules with nitrification ability were formed (Fig. 10.9). After the system reached the steady state,

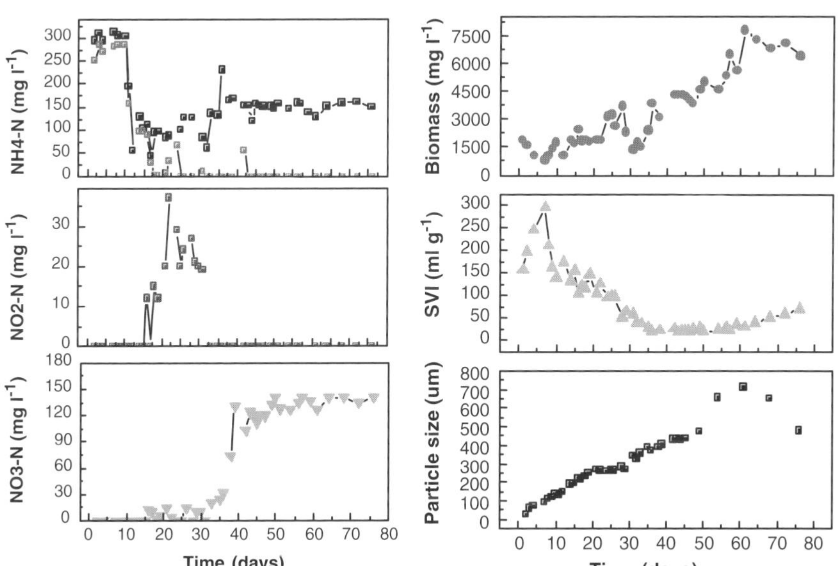

Fig. 10.9. Parameters of growth and nitrification in microbial granules during their cultivation in SBR.

the mean diameter of granules, SVI of granular biomass, and biomass concentration was 0.6 mm, 22 mL g^{-1}, and 7 g L^{-1}, respectively. Nitrite was not detected in the effluent, and ammonium consumption efficiency was very close to 100%. Conversion of the consumed NH_4^+-N to NO_3^--N was 93% and the specific NO_3^--N production rate was 0.12 d^{-1} at the constant NH_4^+-N loading late of 0.9 kg m^{-3} d^{-1}.

Figure 10.10 shows the time course of parameters in one cycle by SBR operation. Complete nitrification could be finished during three hours. pH in one cycle first decreased to the lowest value and then increased to a steady state. This pH changing trend has been confirmed in many batch cultures. Peng (1999) also reported the same pH changing trend during

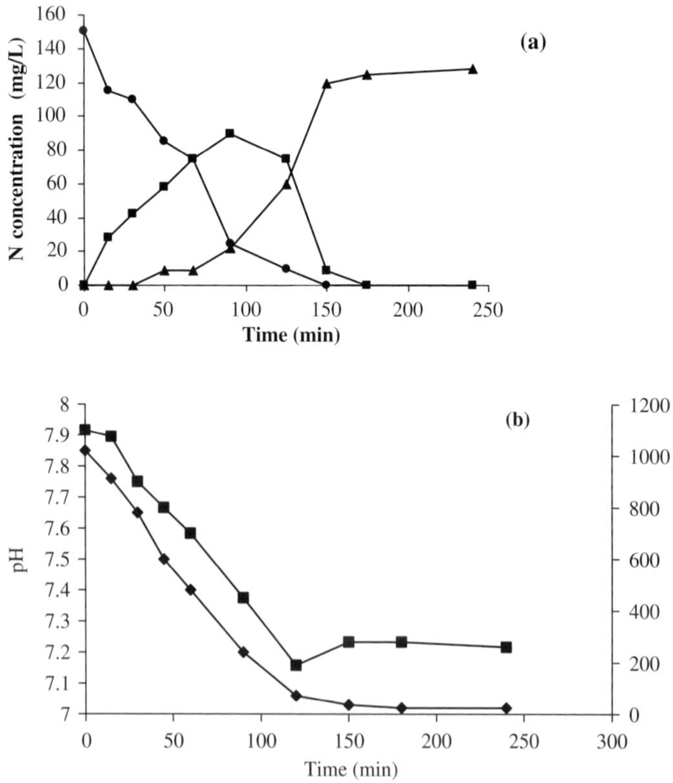

Fig. 10.10. Parameters of nitrification in microbial granules in one cycle of cultivation in SBR. Following parameters are shown: (a) concentrations of ammonium (•), nitrite (■), and nitrate (▲); (b) pH (♦) and alkalinity (■).

nitrification process by nitrifying sludge. This interesting phenomenon could be applied to monitor if there was complete nitrification, which was much easier than the analysis of nitrogen.

Formation of Phenol-degrading Granules from Acetate-fed Granules

Phenol is a major environmental pollutant, and phenol concentrations of up to 10,000 mg L^{-1} have been reported in many industrial wastewaters (Fedorak and Hrudey, 1988). Phenol removal by biological methods is generally preferred to physico-chemical methods because of lower costs and the possibility of complete mineralization. However, phenol-containing wastewater is difficult to treat as microbial activity can be inhibited due to the toxicity exerted by high concentrations of the substrate itself. Although biological treatment of phenol wastewater can be achieved with conventional activated sludge systems, such systems have been known to break down because of fluctuations in phenol loads or because of exposure to high phenol loading rates in excess of 1 kg phenol m^{-3} d^{-1} (Watanabe et al., 1999).

The inhibitory difficulties associated with high-strength phenolic wastewaters can be overcome by strategies such as bioaugmentation (Watanabe et al., 2002) and cell immobilization. Aerobic granules are self-immobilized aggregates of microorganisms and organic and inorganic matter held together by a matrix of extracellular polymers (Morgenroth et al., 1997; Beun et al., 1999; Moy et al., 2002). Aerobic granules have a strong, compact microbial structure, good settling ability, and high biomass retention. Aerobic granules are typically cultivated by using activated sludge as a starting inoculum. However, activated sludge might not be suitable for direct inoculation into a reactor that has a high input of chemical toxicity. We previously reported the successful cultivation of aerobic phenol-degrading granules (Jiang et al., 2002) where the microbial inoculum was municipal activated sludge seed that was first conditioned by incubation with phenol for a period of two months. Such long conditioning times might pose a problem in deploying aerobic granules for field application. One solution might be to use a better inoculum.

One objective of the current study was the comparison between unconditioned activated sludge and aerobic acetate-fed granules as a microbial inoculum for treatment of wastewaters with high phenol concentrations. Because the microbial community in the granules contain a high diversity of microorganisms, we hypothesize that granules possess enough physiological traits and a reservoir of functional responses to make them ideal candidates for use as a starting seed to rapidly produce stable granules that can efficiently degrade phenol. Moreover, compared to activated sludge flocs, the compact structure of the acetate-fed granules should provide better protection against phenol toxicity. This work should contribute to a practical understanding of how aerobic granulation technology can be targeted at industrial wastewaters containing high concentrations of toxic chemicals. Tay et al. (2005) applied acetate-fed granules as a starting seed for the development of phenol-degrading granules. Stable phenol-degrading granules were developed within one week after starting the reactor with acetate-fed granules as starter culture.

Activated sludge and acetate-fed granules were used as microbial inoculum to start-up two sequencing batch reactors for phenol biodegradation. The reactors were operated in 4 h cycles at a phenol loading of 1.8 kg m^{-3} d^{-1}. The biomass in R1 failed to remove phenol and completely washed out after four days. R2 experienced difficulty in removing phenol initially, but the biomass acclimated quickly and effluent phenol concentrations declined to 0.3 mg L^{-1} from day 3. The acetate-fed granules were covered with bacterial rods, but filamentous bacteria with sheaths, presumably to shield against toxicity, quickly emerged as the dominant morphotype upon phenol exposure. Bacterial adaptation to phenol also took the form of modifications in enzyme activity and increased production of extracellular polymers. 16S rRNA gene fingerprints revealed a slight decrease in bacterial diversity from day 0 to day 3 in R1, prior to process failure. In R2, a clear shift in community structure was observed as the seed evolved into phenol-degrading granules without losing species richness. The results highlight the effectiveness of granules over activated sludge as seed for reactors treating toxic wastewaters.

Reactors R1 and R2 were operated by feeding with phenol as sole carbon and energy source. A constant loading rate of 1.8 kg phenol m^{-3} day^{-1} was maintained, corresponding to an influent phenol concentration of 600 mg L^{-1}. By operating at a volumetric exchange ratio of 50%, this was diluted to a phenol concentration of 300 mg L^{-1} in the reactor. The biomass concentration in R1 dropped sharply from 3.7 to 0.2 g L^{-1} within the first two days, and this was accompanied by rapid system failure. R1 was unable to biologically remove the phenol and the phenol concentration in the effluent rose to 570 mg L^{-1} on day 2. Consequently, the biomass was completely washed out of R1 by day 4. In contrast, R2 showed good biomass retention, and the biomass concentration stabilized at 4.6 g L^{-1} within two weeks after start-up. SVI values showed a gradual increase in the first three weeks but stabilized below 80 mL g^{-1} towards the end of the reactor operation, indicating that the granules continued to possess good settleability. R2 experienced some initial difficulty in removing phenol, as phenol concentrations in the effluent increased from 300 mg L^{-1} to 500 mg L^{-1} in the first two days of reactor operation. However, this lag lasted briefly, and phenol concentrations in the effluent rapidly declined to stabilize at 0.3 mg L^{-1} from day 3. Low specific mineralization activities were initially recorded for R2 biomass (reaching approximately 10 mg CO_2 g VSS^{-1} on day 3) but improved quickly to stay above 20 mg CO_2 g VSS^{-1} beyond day 11.

Figure 10.11 shows the morphological changes in the acetate-fed granules upon exposure to phenol. The acetate-fed granules that were used to seed reactor R2 consisted of lumps of microcolonies agglomerated together. The granule surface initially consisted mostly of bacterial rods embedded in an extracellular polymeric matrix.

Filamentous bacteria started to emerge in isolated pockets on the granule surface on day 3, and became the dominant morphotype by day 15. These filamentous bacteria had long, straight, or curved filaments with round-ended or rod-shaped cells within a clear tight-fitting sheath, contained cell septa with indentations, exhibited false branching, stained gram-negative and Neisser-negative, and did not contain any sulfur granules (Fig. 10.12). This morphological description is consistent with that of *Sphaerotilus natans* (Jenkins et al., 1993).

Figure 10.13 shows representative denaturing gradient gel electrophoresis (DGGE) profiles of the biomass in R1 on days 0 and 3 and in R2 on days 0, 3, 15, and 30.

Identical fingerprint patterns were obtained for replicate samples. R1 exhibited a slight decrease in community diversity from day 0 to day 3, just before the onset of process failure, as evidenced by a decrease in SDI from 1.27 to 1.15. On the other hand, community diversity was slightly

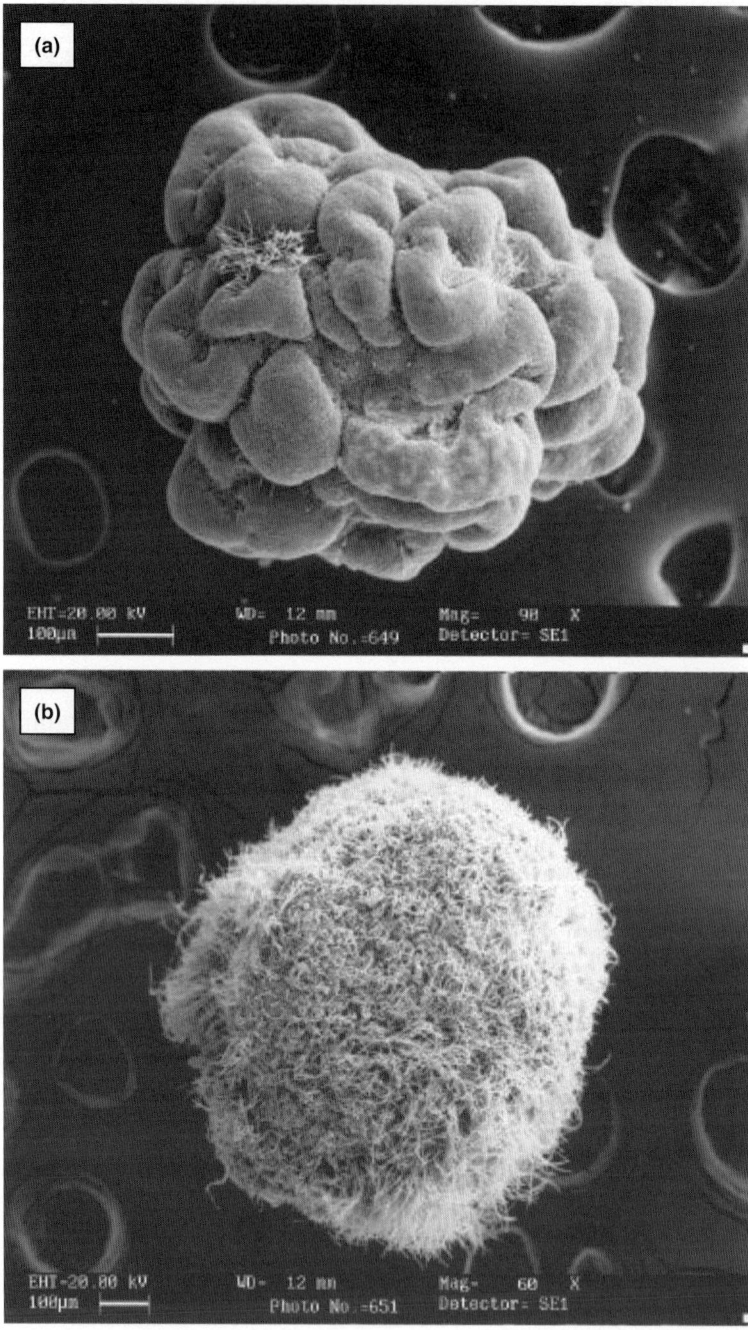

Fig. 10.11. Scanning electron microscopy images of granules on day 3 (a) and day 15 (b).

Fig. 10.12. Light microscopy image of sheath bacteria on surface of phenol-degrading granule. Scale bar is 10 μm long. (See Color Plate Section before the Index.)

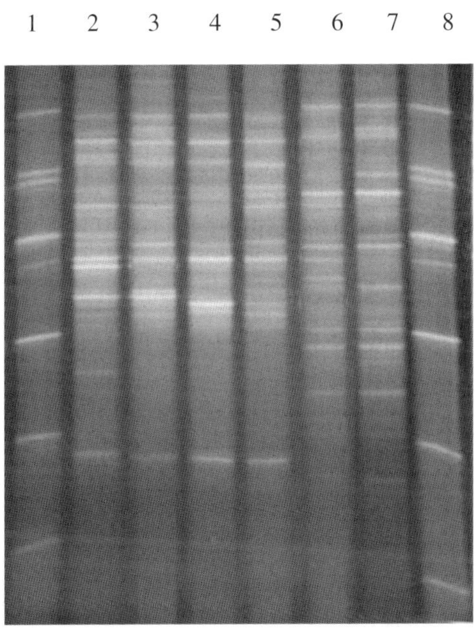

Fig. 10.13. DGGE profiles of R1 and R2 using partial bacterial 16S rRNA gene fragments. Lanes: 1, migration standards; 2, R1 biomass on day 0; 3, R1 biomass on day 3; 4, R2 biomass on day 0; 5, R2 biomass on day 3; 6, R2 biomass on day 15; 7, R2 biomass on day 30; 8, migration standards. (See Color Plate Section before the Index.)

lower in R2 compared to R1, but no significant declines in community diversity were observed for R2, and SDI values ranged from 1.12 to 1.18 over the entire duration of reactor operation. EI values were above 0.97 in both R1 and R2, and indicated reasonably even distribution of different species within each community. Cluster analysis of the DGGE data showed that the community structures in R2 on days 0 and 3 was closer in similarity to the community structures in R1 on days 0 and 3 than to the community structures in R2 on days 15 and 30. The community structures in the acetate-fed granule seed in R2 and in the activated sludge in R1 shared a similarity of 78%. However, the community structures of day 0 and day 30 granules showed less than 55% similarity, and clearly revealed a marked change in the bacterial community in the R2 granules as they adapted to the phenol input.

Seeds for Phenol-degrading Granules

The activated sludge seed failed to maintain an adequate level of biomass within reactor R1 and could not acclimate quickly enough to allow phenol-degrading microorganisms to multiply and remove the phenol. As a consequence, phenol rapidly accumulated in the reactor and the biomass was completely washed out of R1 within four days after start-up. On the other hand, the use of acetate-fed granules as a starting seed resulted in the development of stable phenol-degrading granules with good settling ability, good biomass retention and good metabolic activity, as evidenced by the low SVI values, stable biomass concentrations and nearly complete phenol removal. Although there was a slight lag in the ability of the acetate-fed granules to degrade phenol initially, the compact structure of the acetate-fed granules likely provided the microorganisms with adequate protection against phenol toxicity and minimized sludge washout, thus allowing the buildup of a critical population of phenol-degrading microorganisms such as the filamentous bacteria observed in Fig. 10.12. The different DGGE banding patterns in the steady-state phenol-degrading granules compared to the acetate-fed granule seed indicated that some form of community restructuring had taken place. The granules quickly acclimated to the phenol load and achieved complete phenol removal three days after start-up. The granules stabilized within two weeks after start-up, with little change in biomass concentration, phenol removal, and specific mineralization activity.

Exposure of the granules to phenol triggered a two-fold increase in ECP content two weeks into the reactor operation. This was associated with an increase in PN production and the proliferation of sheath bacteria on the granule surface. ECPs are the construction materials for microbial aggregates and are responsible for their structural integrity. They also serve a protective function and are known to form a shield against the adverse influences of the external environment by acting as a diffusion limitation barrier to delay or prevent toxicants from reaching the microorganisms (Wingender et al., 1999). PS and PN play different roles within the ECP matrix, the stability of which depends on the interactions between PS and PN and the other macromolecules present (Flemming and Wingender, 2001; Sutherland, 2001). A similar preferential production of PN over PS in ECPs has also been observed in other biofilms and granules exposed to high phenol concentrations (Fang et al., 2002; Jiang et al., 2004a).

The propagation of filamentous bacteria is generally thought to be favored by low nutrient or low oxygen conditions (Jenkins et al., 1993). According to the kinetic selection theory, filamentous bacteria are considered to be slow-growing microorganisms with maximum growth rates (μ_{max}) and affinity constants (K_S) lower than floc-forming bacteria (Martins et al., 2004). In systems where the substrate concentration is high, like in plug-flow reactors and the SBR system used in the current study, the filamentous bacteria should be suppressed since their growth rate is expected to be lower than that for the floc-forming bacteria. Therefore, the emergence and eventual dominance of filamentous bacteria in the granules in R2 was an interesting and unexpected development. Even under the high concentrations of phenol substrate in R2, filamentous bacteria were the dominant bacterial morphotype residing on the granule surface. Although stresses such as substrate overloads are known to induce the proliferation of filamentous bacteria, this is thought to be the result of the oxygen shortage induced by the transient substrate overload rather than the massive substrate input itself (Pernelle et al., 2001). However, oxygen deficiency is not expected to be a problem in the current study because of the high aeration rates employed in R2. The precise reasons for the dominance of the filamentous bacteria in the phenol-degrading granules must be linked to their ability to compete in a highly toxic environment. The tolerance to phenol is probably due to the presence of a sheath that is composed of proteins, polysaccharides, and lipids, which would serve as a protective barrier against phenol toxicity. This notion is corroborated by surveys of aquatic biofilms in highly polluted rivers where the dominance of filamentous bacteria was associated with their ability to tolerate high concentrations of pollutants and metals in the rivers (Brummer et al., 2003). Chlorine decay assays also lend support to this idea, as sheathed *Sphaerotilus natans* are known to be several-fold more resistant to chlorination than the floc-forming but sheathless *Acinetobacter anitratus* (Caravelli et al., 2003).

Activated sludge-derived granules were a more appropriate inoculum than activated sludge for the development of phenol-degrading granules. The use of activated sludge resulted in system failure. On the other hand, the compact structure of the granules afforded sufficient protection against phenol toxicity and minimized sludge washout, thus facilitating a rapid microbial acclimation towards phenol biodegradation. This strategy of using granules as a microbial inoculum has practical implications for starting up aerobic granulation systems treating wastewaters containing high concentrations of toxic chemicals.

References

Arrojo, B., Mosquera-Corral, A., Garrido, J.M., & Mendez, R. (2004). Aerobic granulation with industrial wastewater in sequencing batch reactors. *Water Res.*, *38*, 3389–3399.

Bergsma-Vlami, M., Prins, M.E., Staats, M., & Raaijmakers, J.M. (2005). Assessment of genotypic diversity of antibiotic-producing pseudomonas species in the rhizosphere by denaturing gradient gel electrophoresis. *Appl. Environ. Microbiol.*, *71*, 993–1003.

Beun, J.J., Hendriks, A., van Loosdrecht, M.C.M., Morgenroth, E., Wilderer, P.A., & Heijnen, J.J. (1999). Aerobic granulation in a sequencing batch reactor. *Water Res.*, *33*, 2283–2290.

Beun, J.J., van Loosdrecht, M.C.M., & Heijnen, J.J. (2000). Aerobic granulation. *Water Sci. Technol.*, *41*, 41–48.

Bossier, P., & Verstraete, W. (1996). Triggers for microbial aggregation in activated sludge? *Appl. Environ. Microbiol.*, *45*, 1–6.

Brummer, I.H.M., Felske, A., & Wagner-Dobler, I. (2003). Diversity and seasonal variability of beta-proteobacteria in biofilms of polluted rivers: analysis by temperature gradient gel electrophoresis and cloning. *Appl. Environ. Microbiol.*, *69*, 4463–4473.

Calleja, G.B. (1984). *Microbial Aggregation*, CRC Press, Inc., USA.

Caravelli, A., Contreras, E.M., Giannuzzi, L., & Zaritzky, N. (2003). Modeling of chlorine effect on floc forming and filamentous micro-organisms of activated sludges. *Water Res.*, *37*, 2097–2105.

Dabert, P., Delgenès, J.-P., & Godon, J.-J. (2005). Monitoring the impact of bioaugmentation on the start up of biological phosphorus removal in a laboratory scale activated sludge ecosystem. *Appl. Microbiol. Biotechnol.*, *66*, 575–588.

de Bruin, L., de Kreuk, M.K., van der Roest, H.F.R., Uijterlinde, C., & van Loosdrecht, M.C.M. (2004). Aerobic granular sludge technology: an alternative to activated sludge? *Water Sci. Technol.*, *49*, 1–7.

de Kreuk, M.K., & van Loosdrecht M.C.M. (2004). Selection of slow growing organisms as a means for improving aerobic granular sludge stability. *Water Sci. Technol.*, *49*, 9–17.

de Kreuk, M.K., de Bruin, L.M.M., & van Loosdrecht, M.C.M. (2005). Aerobic granular sludge - from ideal to pilot plant. *Aerobic granular sludge* (eds. Bathe, S., de Kreuk, M., McSwain, B.S., & Schwarzenbeck, N.), IWA publishing, London, 111–124.

Etterer, T., & Wilderer, P.A. (2001). Generation and properties of aerobic granular sludge. *Water Sci. Technol.*, *43*, 19–26.

Fang, H.H.P., Xu, L.C., & Chan, K.Y. (2002). Effects of toxic metals and chemicals on biofilm and biocorrosion. *Water Res.*, *36*, 4709–4716.

Fedorak, P.M., & Hrudey, S.E. (1988). Anaerobic degradation of phenolic compounds with application to treatment of industrial waste waters. *Biotreatment Systems, Vol. 1* (ed. Wise, D.L.), CRC Press, Boca Raton, FL, 170–212.

Flemming, H.C., & Wingender, J. (2001). Relevance of microbial extracellular polymeric substances (EPSs) – part I: structural and ecological aspects. *Water Sci. Technol.*, *43* (6), 1–8.

Glancer, M., Ban, S., Soljan, V., & Rpascik, I. (1994). Upgrading of waste-water treatment plants for the biological nitrogen elimination by the injection of on-line prepared mixed cultures. *Water Sci. Technol.*, *29*, 129–138.

Ivanov, V., Tay, J.-H., Tay, S.T.-L., & Jiang, H.-L. (2004). Removal of micro-particles by microbial granules used for aerobic wastewater treatment. *Water Sci. Technol.*, *50*, 147–154.

Ivanov, V., Tay, J.-H., Liu, Q.-S., Wang, X.-H., Wang, Z-.W., Maszenan, B.A.M., Yi, S., Zhuang, W.-Q., Liu, Y.-Q., Pan, S., & Tay, S.T.-L. (2005). Microstructural optimization of wastewater treatment by aerobic granular sludge *Aerobic Granular Sludge* (eds. Bathe, S., de Kreuk, M., McSwain, B., & Schwarzenbeck, N.), IWA Publishing, London, 43–52.

Jenkins, D., Richard, M.G., & Daigger, G.T. (1993). *Manual on the Causes and Control of Activated Sludge Bulking and Foaming*. (2nd edn.). Lewis Publ. Boca Raton, F.L., USA.

Jiang, H.L., Tay, J.-H., & Tay, S.T.-L. (2002). Aggregation of immobilized activated sludge cells into aerobically grown microbial granules for the aerobic biodegradation of phenol. *Lett. Appl. Microbiol.*, *35*, 439–445.

Jiang, H.L., Tay, J.-H., & Tay, S.T.-L. (2004a). Changes in structure, activity and metabolism of aerobic granules as a microbial response to high phenol loading. *Appl. Microbiol. Biotechnol.*, *63*, 602–608.

Jiang, H.L., Tay, J.-H., Maszenan, A.M., & Tay, S.T.-L. (2004b). Bacterial diversity and function of aerobic granules engineered in a sequencing batch reactor for phenol degradation. *Appl. Environ. Microbiol.*, *70*, 6767–6775.

Jorand, R., Zartarian, F., Thomas, F., Block, J.C., Bottero, J.Y., Villemin, G., Urbain, V., & Manem, J. (1995). Chemical and structural (2D) linkage between bacteria within activated sludge flocs. *Water Res.*, *29*, 1630–1647.

Li, D., & Ganczarczyk, J.J. (1990). Structure of activated sludge flocs. *Biotechnol. Bioeng.*, *35*.

Limbergen, H.V., Top, E.M., & Verstraete, W. (1998). Bioaugmentation in activated sludge:current features and future perspectives. *Appl. Microbiol. Biotechnol.*, *50*, 16–23.

Liu, Y., & Tay, J.-H. (2002). The essential role of hydrodynamic shear force in the formation of biofilm and granular sludge. *Water Res.*, *36*, 1653–1665.

Liu, Y., & Tay, J.-H. (2004). State of the art of biogranulation technology for wastewater treatment. *Biotechnol. Adv.*, *22*, 533–563.

Martins, A.M.P., Pagilla, K., Heijnen, J.J., & van Loosdrecht, M.C.M. (2004). Filamentous bulking sludge – a critical review. *Water Res.*, *38*, 793–817.

Morgenroth, E., Sherden, T., van Loosdrecht, M.C.M., Heijnen, J.J., & Wilderer, P.A. (1997). Aerobic granular sludge in a sequencing batch reactor. *Water Res.*, *31*, 3191–3194.

Moy, B.Y.P., Tay, J.-H., Toh, S.K., Liu, Y., & Tay, S.T.-L. (2002). High organic loading influences the physical characteristics of aerobic granules. *Lett. Appl. Microbiol.*, *34*, 407–412.

Nicolella, C., van Loosdrecht, M.C., & Heijnen, S.J. (2000). Particle-based biofilm reactor technology. *Trends Biotechnol.*, *18*, 312–320.

Peng, D.C., Bernet, N., Delgenes, J.-P, & Moletta, R. (1999). Aerobic granular sludge – a case report. *Water Res.*, *33*, 890–893.

Pernelle, J.J., Gaval, G., Cotteux, E., & Duchene, P. (2001). Influence of transient substrate overloads on the proliferation of filamentous bacterial populations in an activated sludge pilot plant. *Water Res.*, *35*, 129–134.

Qin, L., Liu, Y., & Tay, J.-H. (2004). Effect of settling time on aerobic granulation in sequencing batch reactor. *Biochem. Eng. J.*, *21*, 47–52.

Schwarzenbeck, N., Borges, J.M., & Wilderer, P.A. (2005). Treatment of dairy effluents in an aerobic granular sludge sequencing batch reactor. *Appl. Microbiol. Biotechnol.*, *66*, 711–718.

Snidaro, D., Zartarian, F., Jorand, F., Bottero, J.-Y., Block, J.-C., & Manem, J. (1997). Characterisation of activated sludge flocs structure. *Water Sci. Technol.*, *36*, 313–320.

Sutherland, I.W. (2001). Biofilm exopolysaccharides: a strong and sticky framework. *Microbiol.*, *147*, 3–9.

Tay, J.-H., Liu, Q.S., & Liu, Y. (2001a). Microscopic observation of aerobic granulation in sequential aerobic sludge blanket reactor. *J. Appl. Microbiol.*, *91*, 168–175.

Tay, J.-H., Liu, Q.-S., & Liu, Y. (2001b). The effects of shear force on the formation, structure and metabolism of aerobic granules. *Appl. Microbiol. Biotechnol.*, *57*, 227–233.

Tay, J.-H., Liu, Q.-S., & Liu, Y. (2001c). Microscopic observation of aerobic granulation in sequential aerobic sludge blanket reactor. *J. Appl. Microbiol.*, *90*, 1–8.

Tay, J.-H., Ivanov, V., Pan, S., & Tay, S.T.-L. (2002a). Specific layers in aerobically grown microbial granules. *Lett. Appl. Microbiol.*, *34*, 254–257.

Tay, S.T.-L., Ivanov, V., Yi, S., Zhuang, W.-Q., & Tay, J.-H. (2002b). Presence of anaerobic *Bacteroides* in aerobically grown microbial granules. *Microb. Ecol.*, *44*, 278–285.

Tay, J.-H., Pan, S., Tay, S.T.-L., Ivanov, V., & Liu, Y. (2003a). The effect of organic loading rate on the aerobic granulation: the development of shear force theory. *Water Sci. Technol.*, *47*, 235–240.

Tay, J.-H., Tay, S.T.-L., Ivanov, V., Pan, S., Jiang, H.-L., & Liu, Q.-S. (2003b). Biomass and porosity profiles in microbial granules used for aerobic wastewater treatment. *Lett. Appl. Microbiol.*, *36*, 297–301.

Tay, S.T.-L., Moy, B.Y.-P., Jiang. H.-L., & Tay. J.-H. (2005). Rapid cultivation of stable aerobic phenol-degrading granules using acetate-fed granules as microbial seed. *J. Biotechnol.*, *115*, 387–395.

Watanabe, K., Teramoto, M., & Harayama, S. (1999). An outbreak of nonflocculation catabolic populations caused the breakdown of a phenol-digesting activated sludge process. *Appl. Environ. Microbiol.*, *65*, 2813–2819.

Watanabe, K., Teramoto, M., & Harayama, S. (2002). Stable augmentation of activated sludge with foreign catabolic genes harboured by an indigenous dominant bacterium. *Environ. Microbiol.*, *4*, 577–583.

Wingender, J., Neu, T.R., & Flemming, H.C. (1999). What are bacterial extracellular polymeric substances. *Microbial Extracellular Polymeric Substances* (eds. Wingender, J., Neu, T.R., & Flemming, H.C.), Springer-Verlag, Berlin, 1–19.

Zhu, J.R., & Wilderer, P.A., (2003). Effect of extended idle conditions on structure and activity of granular activated sludge. *Water Res.*, *37*, 2013–2018.

Chapter 11

Biosorption Properties of Aerobic Granules

Yu Liu

Introduction

Heavy metals are often present in a wide variety of industrial wastewater. Heavy metals are non-biodegradable and accumulative in the environment and affect human health when they enter into the food chain. So far, stringent limits on metal concentration have been established due to the relatively high toxicity of heavy metals to environmental receptors. In environmental engineering, more and more research has focused on the removal of heavy metals due to their toxicity to human beings and aquatic life even at relatively low concentrations.

The conventional methods for heavy metal removal from aqueous solution include precipitation with lime or other chemicals, chemical oxidation and reduction, ion-exchange, filtration, electro-chemical treatment, reverse osmosis filtration, evaporative recovery, and solvent extraction. However, when the heavy metal concentrations in the wastewater are low, these processes would have some problems, such as incomplete heavy metal removal, high reagent or energy consumption, generation of toxic sludge or other wastes. Recently, adsorption by activated carbon was applied to remove low-level soluble heavy metals from aqueous solution (Kadirvelu et al., 2001; Mohan and Singh, 2002). While the versatility of activated carbon as adsorbent is obvious in water treatment, it might be costly.

Biosorption is one of the current research focuses looking for inexpensive technology for the removal of heavy metals from aqueous solution. Under this circumstance, a vast array of biomaterials had been tested as biosorbents for heavy metal removal, such as marine algae, fungi, hairy roots of *Thlaspi caertulescens*, wasted activated sludge, digested sludge, and so on (Lodi et al., 1998; Zhou, 1999; Valdman and Leite, 2000; Taniguchi et al., 2002). Most biosorbents used today are suspended microorganisms in forms of bioflocs. One of the major operation problems associated with the suspended flocs is post-separation of biosorbent from the treated effluent. To overcome this drawback, cell immobilization technique is deployed, but the deployment of immobilization procedure is expensive and complex. It should be realized that the disadvantages of conventional biosorbents in the form of bioflocs have seriously limited the application of biological process for the purpose of removal of metals from wastewater.

Aerobic granulation is an innovative biotechnology developed recently (Liu and Tay, 2004a). Aerobic granules are microbial aggregates with strong and compact microbial structure, and settling velocity and density of aerobic granules are much higher than conventional bioflocs (Liu et al., 2005). When selecting appropriate biosorbents for the removal of heavy metals from industrial wastewater, three criteria have to be seriously taken into account, i.e. effectiveness, robustness, and reliability of biosorbents. It appears that the characteristics of aerobic granules may satisfy these requirements for biosorbents (Liu et al., 2002, 2003a,b, 2004b; Xu et al., 2004, 2005). Therefore, this chapter looked into some up-to-date progress of the biosorption of soluble heavy metals by aerobic granules.

Development of a Kinetic Model for Metal Biosorption

Aerobic granules are microbial aggregates with a strong and compact structure. Liu et al. (2003a) investigated the biosorption kinetics of heavy metals by aerobic granules. Figure 11.1 shows the adsorption profiles of cadmium by aerobic granules in the course of batch tests (Liu et al., 2003a). It can be seen that the amount of cadmium adsorbed gradually increased as a function of contact time until a stable level. It had been assumed that functional groups or biopolymers on cell surface would

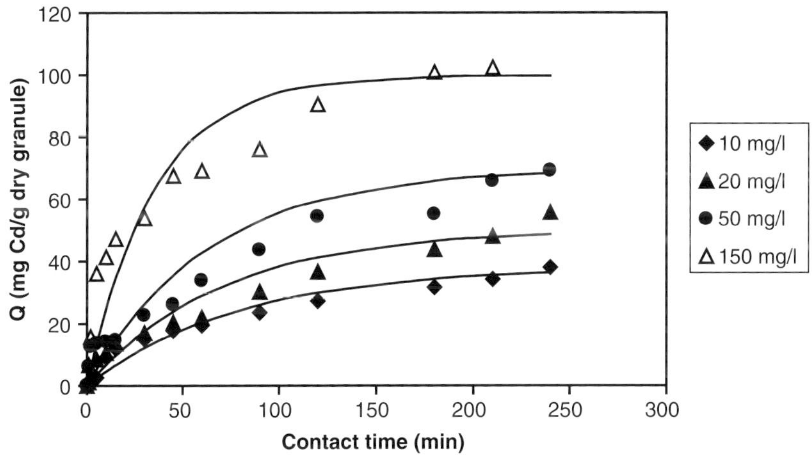

Fig. 11.1. Biosorption profiles of Cd^{2+} at different initial Cd^{2+} concentrations and initial aerobic granules concentration was fixed at 100 mg/l. The model prediction is shown by a solid curve (Liu et al., 2003a).

contribute to the binding of metallic cations by biosorbents, and heavy metal biosorption could be characterized as a physico-chemical process (Guibaud et al., 1999; Jeon et al., 2001; Pethkar et al., 2001) in a way such that

$$S + M \underset{k_2}{\overset{k_1}{\rightleftarrows}} SM \tag{11.1}$$

where S is the available site for metal binding on aerobic granule surface, M is the free metal ions, and SM represents metal ions bound to the site, while k_1 and k_2 are the rate constants for the biosorption and desorption processes, respectively.

As shown in Fig. 11.1, the cadmium adsorption on aerobic granules would be subject to a pseudo first-order reversible process kinetics. In fact, a first-order reaction kinetics had been proposed for biosorptions of Ni^{2+} and Cr^{6+} by *Microcystis* as well as Pb^{2+} by fungal biomass of *Aspergillus niger* (Singh et al., 2001; Wang et al., 2001). Hence, the overall biosorption rate can be written as:

$$-\frac{dC}{dt} = k_1 C - k_2 C_b \tag{11.2}$$

where C is the concentration of the soluble metal ion at time t, C_b is the apparent concentration of the bound metal ions at time t, i.e. the metal ions adsorbed by aerobic granule per unit volume of the solution. A mass balance on metal ions gives

$$C_b = C_0 - C \tag{11.3}$$

where C_0 is the initial concentration of metal ions. When the biosorption process reaches its equilibrium, equation (11.3) becomes

$$C_{be} = C_0 - C_e \tag{11.4}$$

and equation (11.2) reduces to

$$\frac{C_{be}}{C_e} = \frac{k_1}{k_2} \tag{11.5}$$

where C_{be} and C_e is the apparent concentration of the bound metal ions and concentration of free metal ions at biosorption equilibrium, respectively. Combining equations (11.2–11.5) gives

$$-\frac{dC}{dt} = (k_1 + k_2)(C - C_e) \tag{11.6}$$

Integration of equation (11.6) results in

$$\frac{C_0 - C_e}{C - C_e} = e^{kt} \tag{11.7}$$

where $k = k_1 + k_2$ is termed the overall biosorption rate of the metal to aerobic granule. Substituting equations (11.3) and (11.4) into equation (11.7) yields

$$C_b = C_{be}\,\frac{e^{kt} - 1}{e^{kt}} \tag{11.8}$$

Dividing equation (11.8) by aerobic granule concentration yields a general model that describes the metal biosorption on the surfaces of aerobic granules:

$$Q = Q_e(1 - e^{-kt}) \tag{11.9}$$

where Q is the amount of metal ions on aerobic granule surface in terms of milligram metal ions per gram granules at time t, and Q_e is the biosorption capacity at the equilibrium.

Biosorption Kinetics of Various Metals by Aerobic Granules

Biosorption of Cd^{2+} by Aerobic Granules

It can be seen from Figs 11.1 and 11.2 that the proposed kinetic model (equation (11.9)) can provide a satisfactory description for the cadmium biosorption data obtained at various initial cadmium and aerobic granules concentrations, indicated by a coefficient of correlation greater than 0.91. Figure 11.1 shows the biosorption profiles of cadmium at different initial cadmium concentrations (C_0), indicating that about 50% of the amount of adsorbed cadmium at equilibrium was removed in the first 1 h of the test,

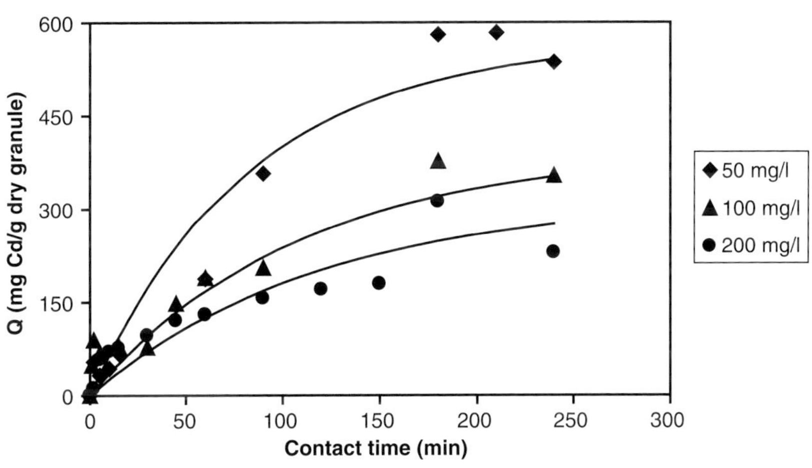

Fig. 11.2. Biosorption profiles of Cd^{2+} at different initial aerobic granules concentrations, while initial Cd^{2+} concentration was kept at 100 mg/l. The model prediction is shown by a solid curve (Liu et al., 2003a).

and the biosorption equilibrium was gradually achieved in 3 h. When the initial granules concentration was constant, the increased Cd^{2+} concentration led to an increase in the Cd^{2+} uptake capacity from 38.2 mg Cd^{2+}/g granules at 10 mg/l Cd^{2+} to 99.8 mg Cd^{2+}/g granules at 150 mg/l Cd^{2+}. These seem to imply that the initial Cd^{2+} concentration has a significant influence on its biosorption by aerobic granules, i.e. biosorption process would slow down when the initial metal concentration is low. Scott and Karanjkar (1992) also reported that when the cadmium concentration increased from 25 to 500 mg/l, the corresponding time required to achieve the equilibrium of biosorption by *Enterobater aerogens* decreased accordingly.

Figure 11.2 shows the biosorption profiles of cadmium at different initial aerobic granules concentrations (X_0), while the initial cadmium concentration was fixed at 100 mg/l. The cadmium biosorption could reach equilibrium within 4 h at all X_0 studied. It should be realized that the cadmium uptake capacity by aerobic granules tend to decrease with increasing initial aerobic granules concentration.

Biosorption of Cu^{2+} by Aerobic Granules

Figure 11.3 shows the biosorption profiles of copper at different initial copper concentrations with a fixed granules concentration of 100 mg/l, while Fig. 11.4 displays the biosorption behaviors of copper at different initial granules concentration with a constant initial copper concentration of 100 mg/l. It can be seen that for both cases, the predictions by equation (11.9) are in good agreement with the experimental data, indicated by a correlation coefficient greater than 0.93.

It appears from Figs 11.3 and 11.4 that the patterns of copper biosorption by aerobic granules are similar to those of cadmium biosorption at various initial metal ions and aerobic granules concentrations. At the constant initial aerobic granules concentration, the increase in the Cu^{2+} concentration led to an increase in the Cu^{2+} biosorption capacity from 8.65 mg Cu^{2+}/g granules at 2 mg/l Cu^{2+} to 29.8 mg Cu^{2+}/g granules at 100 mg/l Cu^{2+}. It seems that the Cu^{2+} biosorption capacity by aerobic granules is related to initial Cu^{2+} concentration, while at a fixed initial Cu^{2+} concentration, an increase of initial aerobic granules concentration would result in a decline in the Cu^{2+} uptake capacity by aerobic granules.

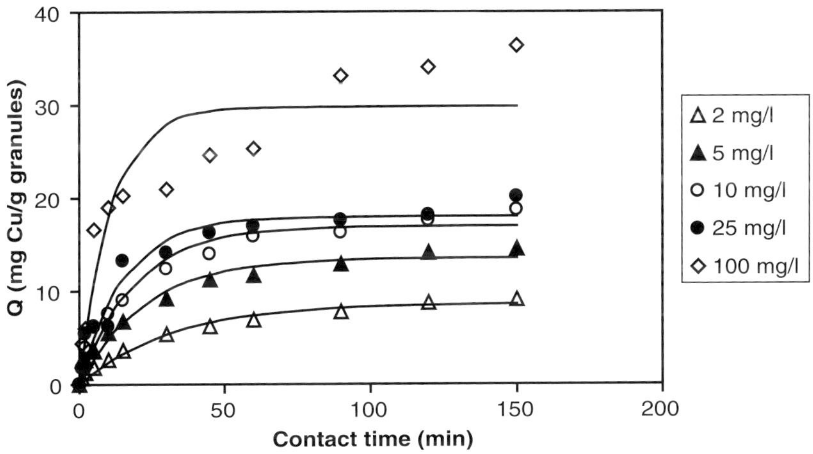

Fig. 11.3. Biosorption profiles of Cu^{2+} at different initial Cu^{2+} concentration with a constant initial aerobic granules concentration of 100 mg/l. The model prediction is shown by a solid curve (Xu, 2006).

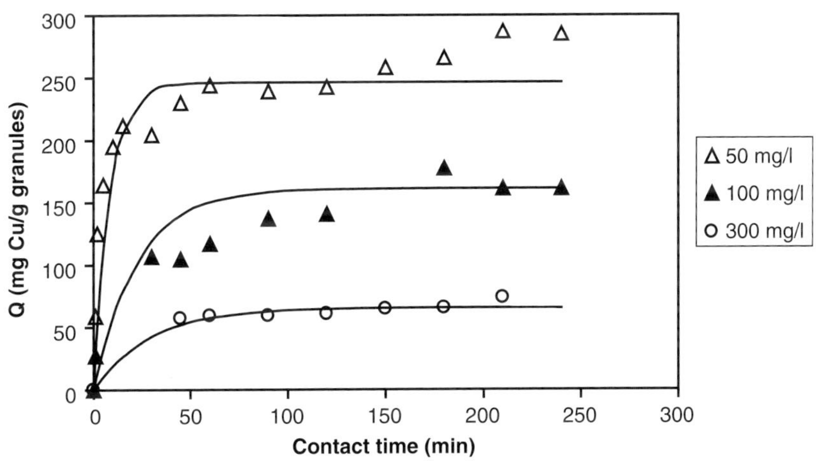

Fig. 11.4. Biosorption profiles of Cu^{2+} at different initial aerobic granules concentrations with a constant initial Cu^{2+} concentration of 100 mg/l. The model prediction is shown by a solid curve (Xu, 2006).

Compared with cadmium biosorption by aerobic granules, the copper removal by aerobic granules is a fast process, e.g. about 50% of the amount of adsorbed copper at equilibrium was removed in the first half hour of contact, and the equilibrium was attained after two hours of contact. Meanwhile, Fig. 11.4 reveals that the uptake of copper at higher initial copper concentrations was faster than that observed at the lower initial copper concentrations, e.g. at an initial copper concentration of 2 mg/l, the time required to attain the biosorption equilibrium was about two hours, however the copper biosorption can reach the equilibrium within one hour at an initial copper concentration of 100 mg/l.

Biosorption of Zn^{2+} by Aerobic Granules

The biosorption profiles of zinc at different initial zinc concentrations (C_0) and aerobic granules concentrations (X_0) were illustrated in Figs 11.5 and 11.6, respectively. It can be seen that the proposed model (equation (11.9)) can fit those biosorption data well, indicated by a correlation coefficient greater than 0.88. It can be seen that the zinc biosorption by aerobic granules could reach the equilibrium within one hour. Compared to the

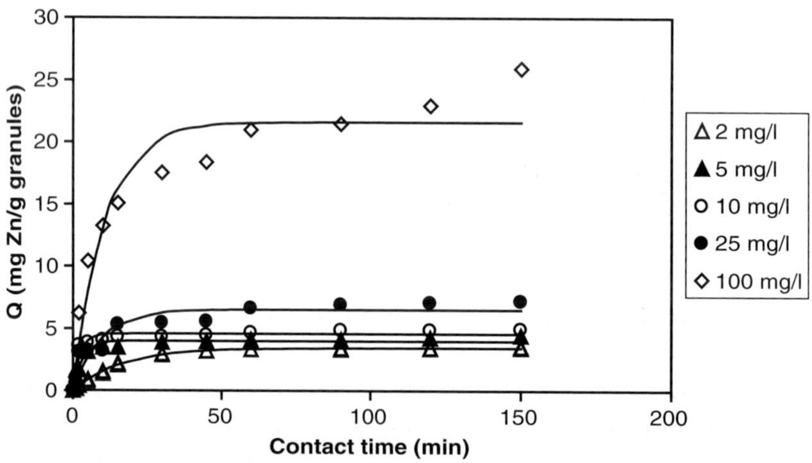

Fig. 11.5. Biosorption profiles of Zn^{2+} at different initial Zn^{2+} concentrations with a constant initial aerobic granules concentration of 100 mg/l. The model prediction is shown by a solid curve (Liu et al., 2002).

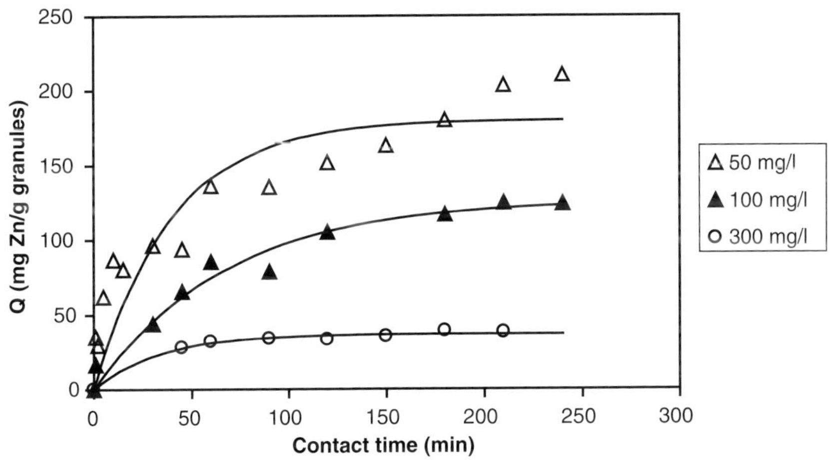

Fig. 11.6. Biosorption profiles of Zn^{2+} at different initial granules concentrations with a constant Zn^{2+} concentration of 100 mg/l. The model prediction is shown by a solid curve (Liu et al., 2002).

biosorption of cadmium and copper, it seems that the zinc uptake by aerobic granules is a faster process.

Effect of Initial Metal Concentration on Biosorption Kinetics

The kinetic parameters involved in equation (11.9) were determined for the biosorption of cadmium, copper, and zinc by aerobic granules (Xu, 2006). This section discussed the effect of initial metal concentration on the biosorption kinetics of metal ions by aerobic granules.

Effect of Initial Metal Concentration on Specific Biosorption Capacity

The relationship between the metal ions uptake capacity and the initial concentration of Cd^{2+}, Cu^{2+}, and Zn^{2+} at the fixed initial granules concentration are illustrated in Figs 11.7–11.9, respectively. It can be seen that the metal biosorption capacity was positively related to the initial metal concentration. These results provide the evidence that the biosorption capacity at equilibrium is highly dependent upon the available

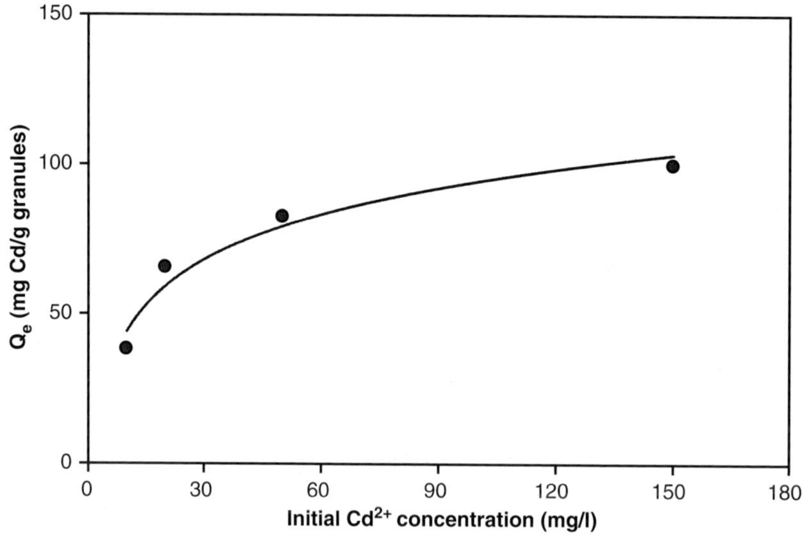

Fig. 11.7. Effect of initial Cd^{2+} concentration on Q_e (Liu et al., 2003a).

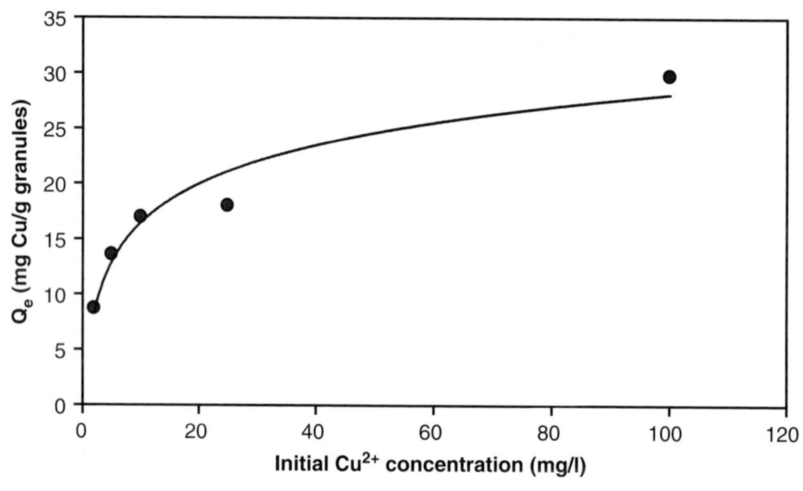

Fig. 11.8. Effect of initial Cu^{2+} concentration on Q_e (Xu, 2006).

binding sites on the aerobic granules surface, which are proportional to the quantity of aerobic granules. Puranik et al. (1999) also reported that the amounts of zinc and lead adsorbed onto *S. cinnamoneum*, *P. chrysogeum*, and *Citrobacter sp.* were increased with the increase of the initial metal concentration.

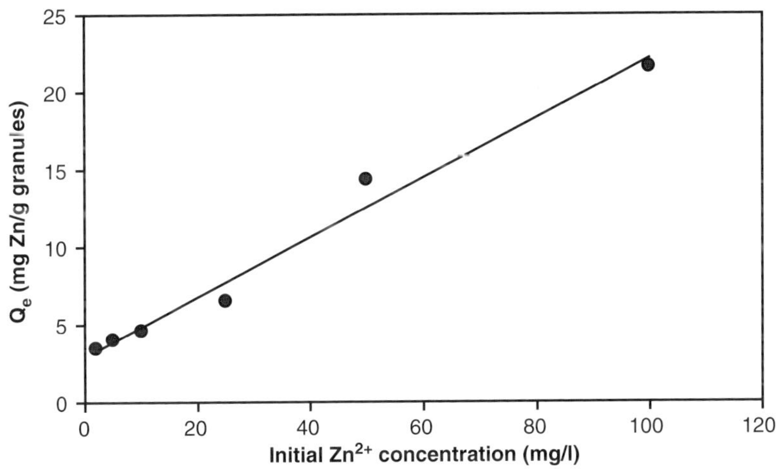

Fig. 11.9. Effect of initial Zn^{2+} concentration on Q_e (Liu et al., 2002).

Effect of Initial Metal Concentration on Overall Biosorption Rate Constant

The relationships between the overall biosorption rate constant (k) and initial metal concentration were shown in Figs 11.10–11.12 for various metal ions (Liu et al., 2003a; Xu, 2006). These results suggest that the increased initial metal concentration favors the biosorption of metal ions, i.e. a faster metal biosorption would occur at higher initial metal concentration.

Effect of Initial Aerobic Granules Concentration on Biosorption Kinetics

Effect of Initial Aerobic Granules Concentration on Specific Biosorption Capacity

The effect of initial aerobic granules concentration on Cd^{2+}, Cu^{2+}, and Zn^{2+} biosorption capacities was shown in the Figs 11.13–11.15, respectively (Xu, 2006). It appears that an increased granules

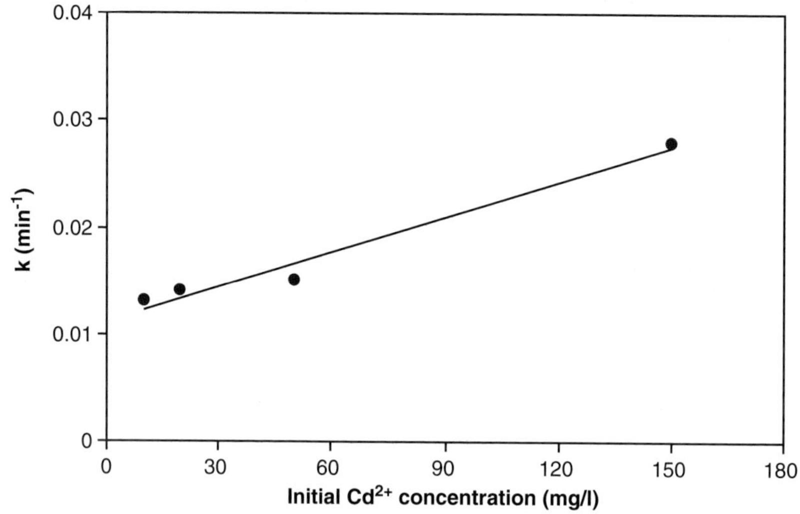

Fig. 11.10. Effect of initial Cd^{2+} concentration on k (Liu et al., 2003a).

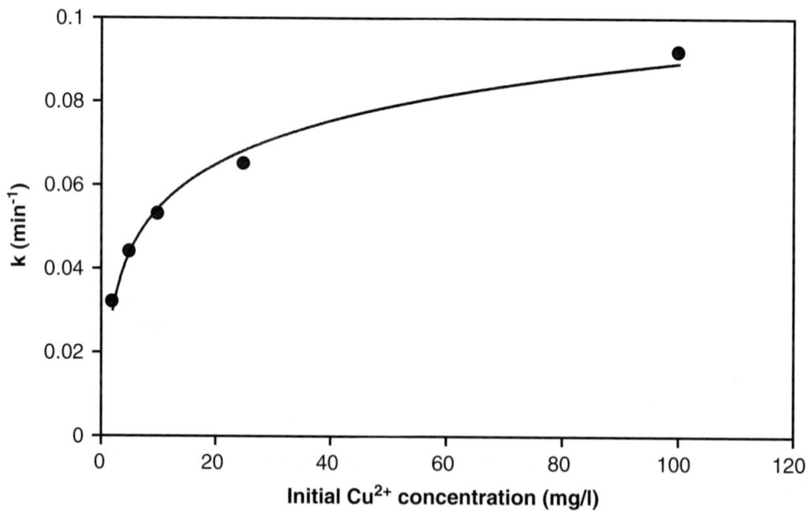

Fig. 11.11. Effect of initial Cu^{2+} concentration on k (Xu, 2006).

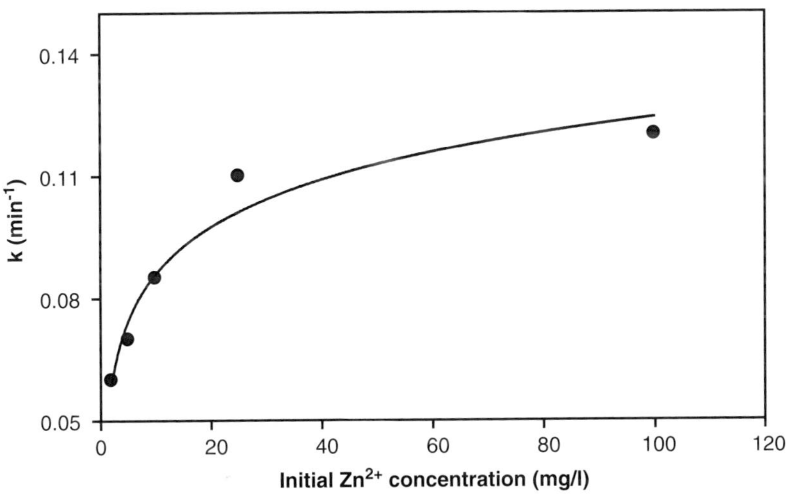

Fig. 11.12. Effect of initial Zn^{2+} concentration on k (Liu et al., 2002).

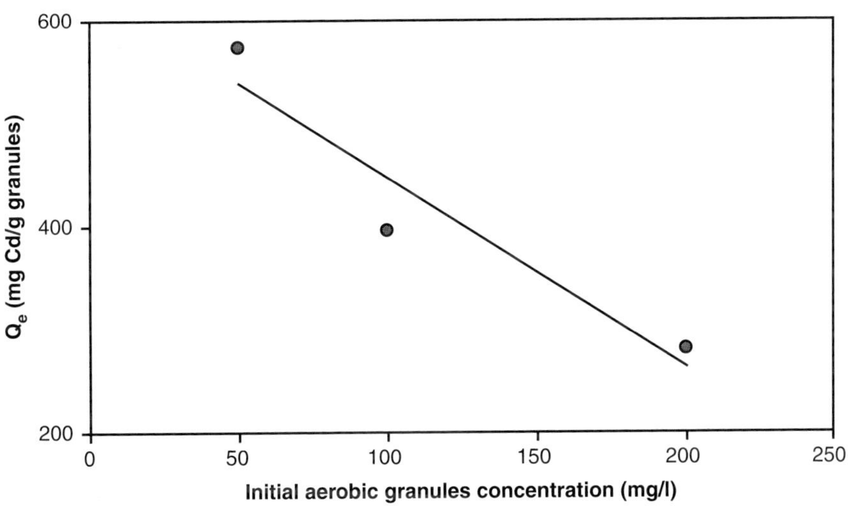

Fig. 11.13. Effect of initial aerobic granules concentration on cadmium biosorption capacity (Xu, 2006).

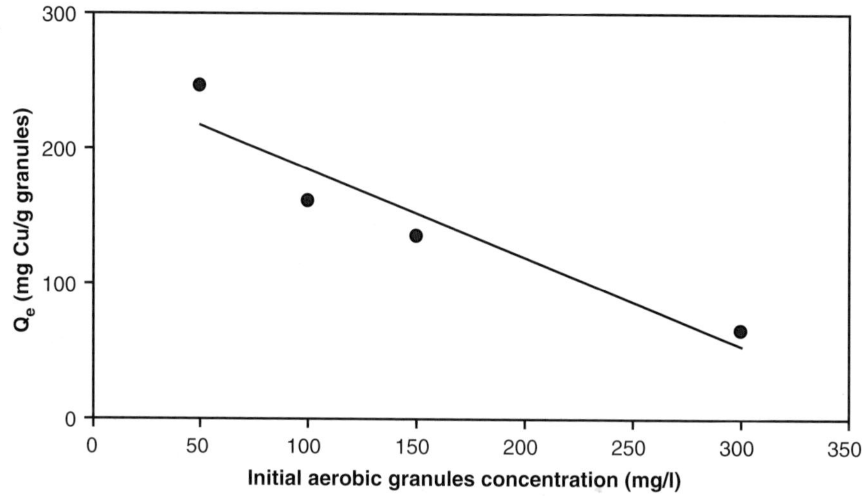

Fig. 11.14. Effect of initial aerobic granules concentration on copper biosorption capacity (Xu, 2006).

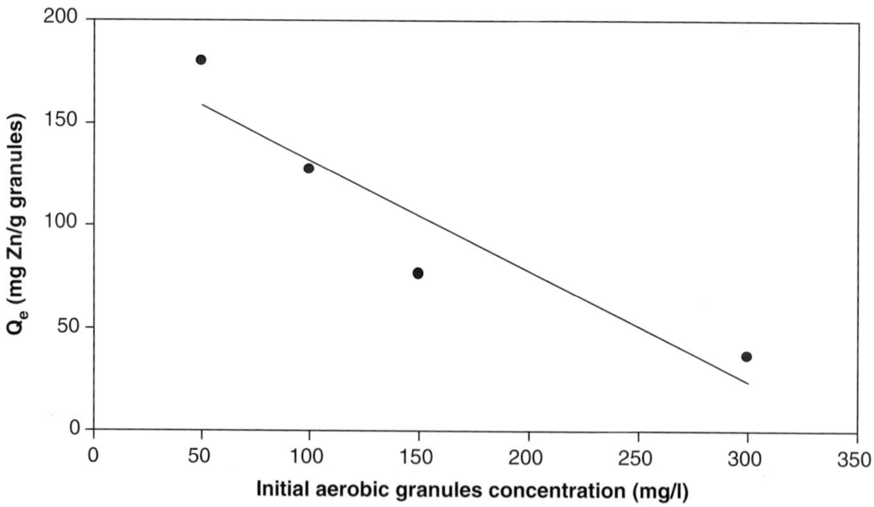

Fig. 11.15. Effect of initial aerobic granules concentration on zinc biosorption capacity (Xu, 2006).

concentration results in a decreased specific biosorption capacity for all three kinds of metal ions.

Effect of Initial Aerobic Granules Concentration on Overall Biosorption Rate Constant

Initial biosorbent and metal concentrations would have a profound effect on the biosorption kinetics (Liu et al., 2002, 2004b). Figure 11.16 shows the relationship between the overall biosorption rate constant (k) of Cd^{2+} and initial aerobic granules concentration (Liu et al., 2003a). The overall biosorption rate constant of Cd^{2+} decreased with the increase of the initial aerobic granules concentration, i.e. k decreased from 0.01 min^{-1} at the initial aerobic granules concentration of 50 mg/l to 0.083 min^{-1} at the initial granules concentration of 200 mg/l.

Figure 11.17 displays the effect of initial granule concentration on k of copper (Xu, 2006). There was a decline of the overall uptake rate constant of copper from 0.12 min^{-1} to 0.045 min^{-1} when the initial granules concentration was increased from 50 mg/l to 100 mg/l. Beyond the initial aerobic granules concentration of 100 mg/l, k of copper seems to be less dependent on the initial aerobic granule concentration. A similar trend was observed in the biosorption of Zn^{2+} by aerobic granules (Fig. 11.18).

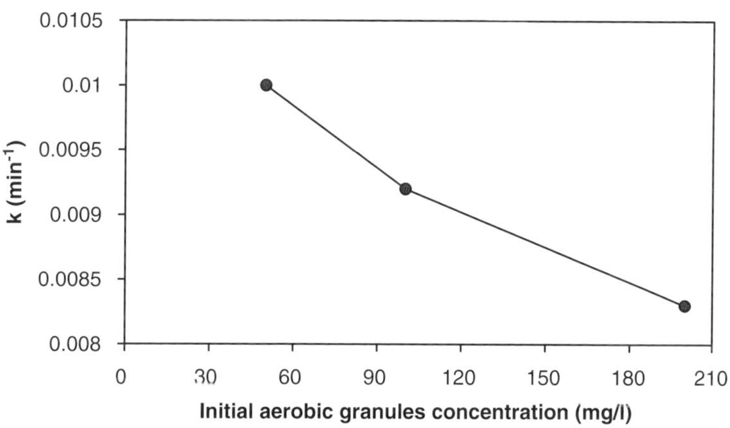

Fig. 11.16. Effect of initial aerobic granules concentration on the cadmium biosorption rate constant (Liu et al., 2003a).

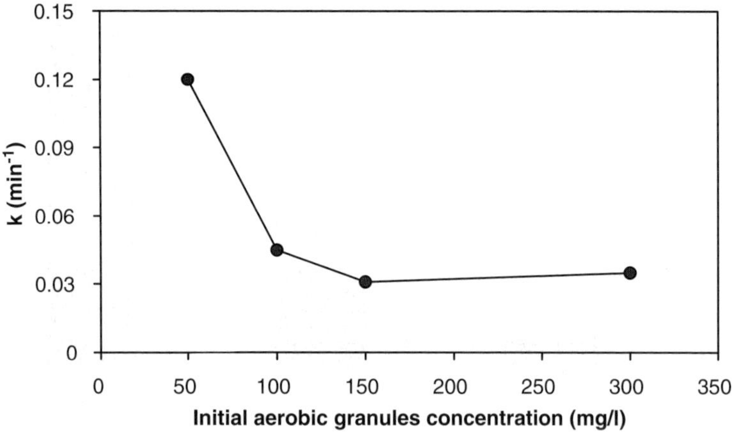

Fig. 11.17. Effect of initial aerobic granules concentration on the copper biosorption rate constant (Xu, 2006).

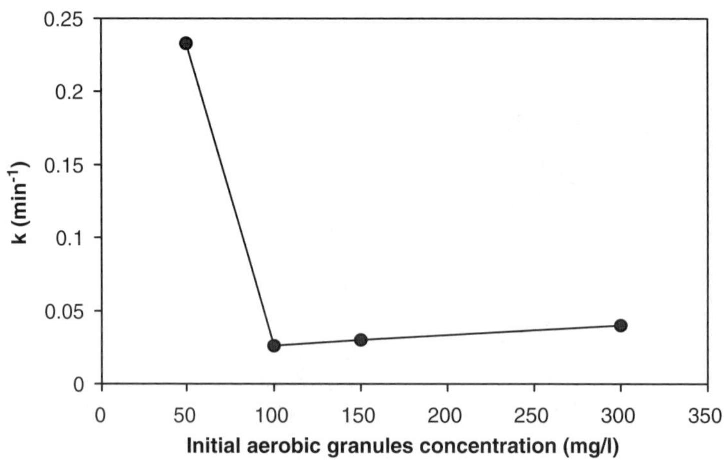

Fig. 11.18. Effect of initial aerobic granules concentration on the zinc biosorption rate constant (Liu et al., 2002).

Comparison of Biosorption Behaviors of Various Metals by Aerobic Granules

Figures 11.19 and 11.20 compare the biosorption of cadmium, copper, and zinc by aerobic granules (Xu, 2006). The biosorption rate constant is in the order of $Cd^{2+} < Cu^{2+} < Zn^{2+}$, indicating that zinc biosorption by aerobic

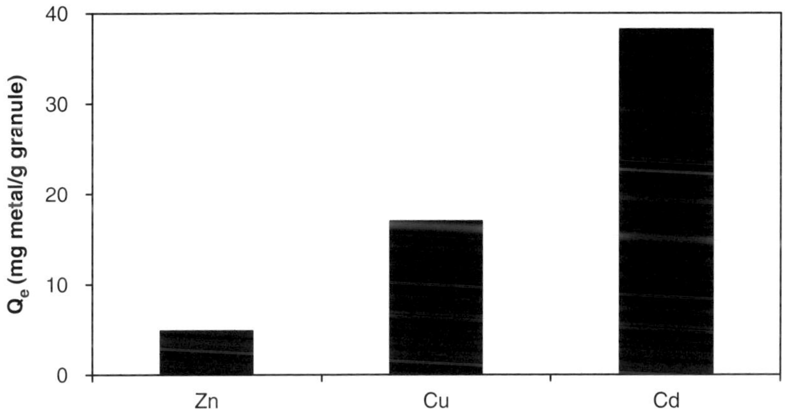

Fig. 11.19. Comparison of various metals on biosorption capacity at $C_0 = 10$ mg/l and $X_0 = 100$ mg/l (Xu, 2006).

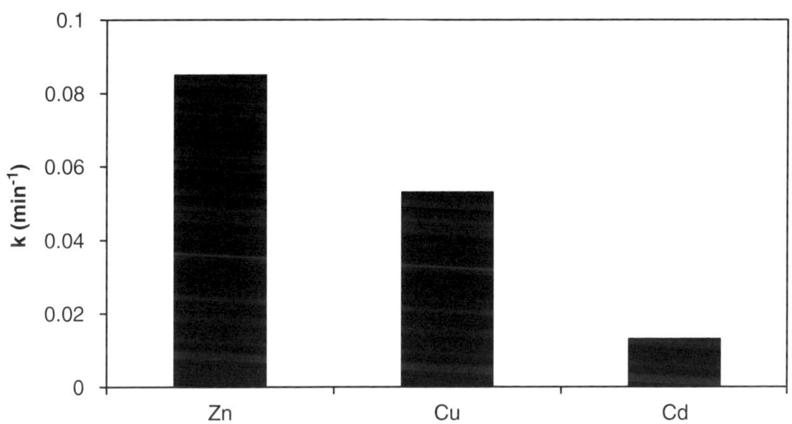

Fig. 11.20. Comparison of various metals on biosorption rate constant at $C_0 = 10$ mg/l and $X_0 = 100$ mg/l (Xu, 2006).

granules was the fastest, while cadmium biosorption by aerobic granules was the slowest. These results could be attributed to the different active sites used for adsorption, which reflect the strength of the metal sorptive bond and the rate of adsorption onto the active sites (Kaewsarn and Yu, 2001). The factors that affect the biosorption preference of biosorbent for different kinds of adsorbates may be related to the characteristics of the binding sites (e.g. functional groups, structure, surface properties, etc.), the properties of the adsorbates (e.g. concentration, ionic size, ionic weight, ionic charge, molecular structure, ionic nature or standard reduction potential, etc.), and the solution chemistry (e.g. pH, ionic strength, etc.) (Aksu and Gülen, 2002).Under comparable test conditions, the amount of metal adsorbed by aerobic granules was subjected to the following order of $Cd^{2+} > Cu^{2+} > Zn^{2+}$ (Fig. 11.19). These results are consistent with the observations by Xu (2002) that the complex stability and binding affinity of metal ions to *Laminaria japonica* is in the order of Zn < Ni < Cu < Cd simply because cadmium usually has higher affinity and stability to bond with carboxyl groups.

A proportional relationship between Q_e and C_0 (Figs 11.7–11.9) indicates that the cadmium, copper, and zinc biosorption on aerobic granule surface could be driven by the concentration gradient of metal at a constant granule concentration. This implies that the driving force for metal biosorption would result from a soluble metal concentration that is higher than the concentration that would be in equilibrium with the amount of metal adsorbed on the aerobic granules. For a constant initial metal concentration, Figs 11.13–11.15 show that Q_e declined as the initial granule concentration increased. Similar phenomena had been reported in studies on lead and zinc uptake by *S. cinnamoneum*, *P. chrysogenum*, and *Citrobacter sp.* (Puranik et al., 1999) and iron (III) and iron (III)-cyanide complex ion uptake by *Rhizopus arrhizus* (Aksu and Gülen, 2002). In the environmental engineering literature, the effect of metal concentration on Q_{eq} was mainly presumed to be due to dilution of the metal with the increased biomass concentration (Taniguchi et al., 2000). It is a reasonable consideration that the number of binding sites to metal on aerobic granules is proportional to the amount of aerobic granules added to the batch tests, i.e. high granule concentration could result in a lower relative metal concentration on the basis of unit mass of aerobic granules.

The biosorption capacity of cadmium, copper, and zinc by aerobic granules was inversely related to their initial aerobic granules concentration (Figs 11.13–11.15), i.e. the metal uptake decreased with the increase of the initial aerobic granules concentration. Othman and Amin (2003) also found that Zn^{2+}, Cu^{2+}, and Mn^{2+} biosorption capacities by a conventional biosorbent decreased from 24.2 to 10.5 mg/g for Zn^{2+},

5 to 2.8 mg/g for Cu^{2+}, and 37.5 to 8.2 mg/g for Mn^{2+} when the biosorbent concentration was increased from 0.5 to 3 g/l. Similar trends were also observed in the biosorption of heavy metals by *Oscillatoria anguistissima*, marine algae, and fungal biomass (Ahuja et al., 1999; Khoo and Ting, 2001).

Figures 11.7–11.9 and 11.13–11.15 indicate that both initial metal and granule concentrations can influence the biosorption capacity of metal at equilibrium, i.e. the biosorption process of metal by aerobic granules cannot be described by C_0 or X_0 alone. When biosorption tests are carried out at a given metal concentration, higher biomass concentration could lower real metal concentration on the basis of unit biomass added. In this case, a concept of relative metal concentration is proposed and defined as the ratio of initial metal concentration to initial granule concentration, i.e. C_0/X_0. This ratio indeed quantifies dilution of metal concentration with the added biomass. The observed relationship between Q_e and C_0/X_0 ratio obtained at various C_0 or X_0 for cadmium, copper, and zinc are presented in Figs 11.21 and 11.22, respectively. It appears that Q_e increases with the increase of C_0/X_0 ratio. These results imply that the individual effects of C_0 and X_0 on the metal biosorption on the surfaces of aerobic granules can be unified by the C_0/X_0 ratio for batch tests initiated at different C_0 and X_0. An important implication of Figs 11.21 and 11.22 is that if C_0 or X_0 is not strictly controlled in batch experiments, the

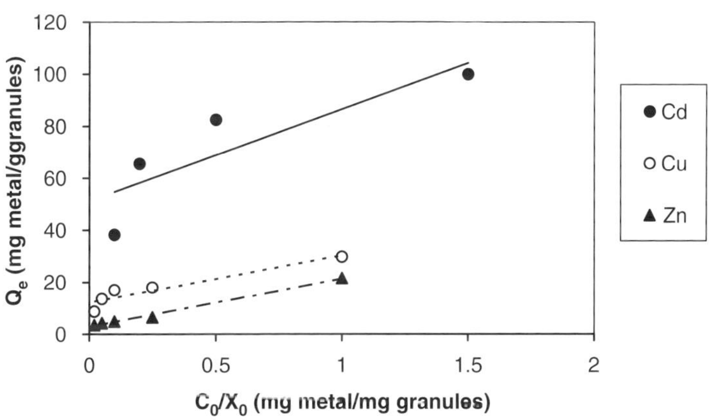

Fig. 11.21. Effect of C_0/X_0 ratio on the heavy metal biosorption capacities at equilibrium (Q_e) at various initial metal concentrations (Xu, 2006).

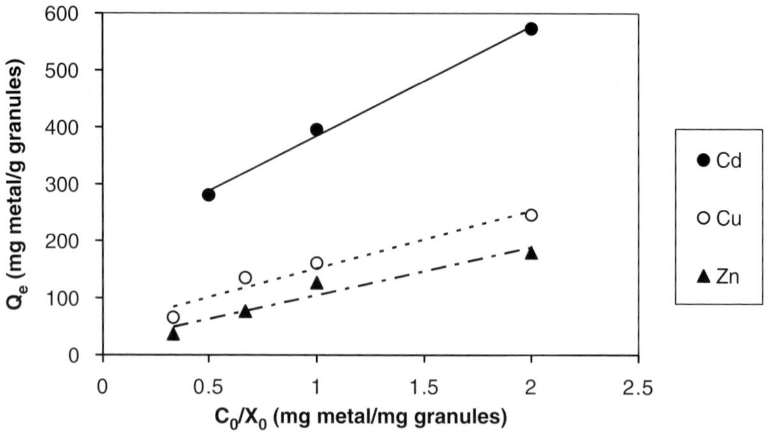

Fig. 11.22. Effect of C_0/X_0 ratio on the heavy metal biosorption capacities at equilibrium (Q_e) at various initial aerobic granules concentrations (Xu, 2006).

C_0/X_0 ratio could better reflect the real driving force for metal biosorption by microorganisms, and provides a unified basis for interpretation of the biosorption data obtained at different initial metal and biomass concentrations.

It should be realized that biosorbents currently used are microbial flocs or dispersed bacteria. One serious operation problem associated with those biosorbents is separation of used biosorbents from the treated effluent. For achieving an efficient solid–liquid separation, an additional settling facility is required. As compared to conventional floc-form biosorbents, aerobic granules have the advantages of compact microbial structure, and excellent settling ability. The settling velocity of the aerobic granules used was 71 m/h, which is 5–8 times higher than that of microbial flocs. In this study, the aerobic granules can be completely separated out from the treated effluent by gravity in one minute. When selecting appropriate biosorbents for the removal of heavy metals from industrial wastewater, three criteria should be seriously taken into account, i.e. effectiveness, robustness, and reliability of biosorbents. It appears that the characteristics of aerobic granules could satisfy these requirements for biosorbents. It can be expected that aerobic granule-based biosorption process is an efficient and cost-effective technology for the removal of heavy metals from industrial wastewater streams.

Summary

Aerobic granules have excellent settleability and high-porosity structure. This chapter shows the feasibility of aerobic granules as a novel type of biosorbent for soluble heavy metal removal from aqueous solution. Future effort would include development of an aerobic granular sludge-based compact biosorber.

References

Ahuja, P., Gupta, R., & Saxena, R.K. (1999). Zn^{2+} biosorption by *Oscillatoria anguistissima*. *Process Biochemistry*, *34*, 77–85.

Aksu, Z., & Gülen, H. (2002). Binary biosorption of iron (III) and iron (III)-cyanide complex ions on *Rhizopus arrhizus*: modeling of synergistic interaction. *Process Biochemistry*, *38*, 161–173.

Guibaud, G., Baudu, M., Dollet, P., Condat, M.L., & Dagot, C. (1999). Role of extracellular polymers in cadmium adsorption by activated sludges. *Environmental Technology*, *20*, 1045–1054.

Jeon, C., Park, J.Y., & Yoo, Y.J. (2001). Biosorption model for binary adsorption sites. *Journal of Microbiology and Biotechnology*, *11*, 781–787.

Kadirvelu, K., Thamaraiselvi, K., & Namasivayam, C. (2001). Removal of heavy metals from industrial wastewaters by adsorption onto activated carbon prepared from an agricultural solid waste. *Bioresource Technology*, *76*, 63–65.

Kaewsarn, P., & Yu, Q.M. (2001). Cadmium(II) removal from aqueous solution by pre-treated biomass of marine alga *Padina sp. Environmental Pollution*, *112*, 209–213.

Khoo, K.M., & Ting, Y.P. (2001). Biosorption of gold by immobilized fungal biomass. *Biochemical Engineering Journal*, *8*, 51–59.

Liu, Y., Yang, S.F., Tan, S.F., Lin, Y.M., & Tay, J.H. (2002). Aerobic granules: a novel zinc biosorbent. *Letters in Applied Microbiology*, *35*, 548–551.

Liu, Y., Yang, S.F., Xu, H., Woon, K.H., Lin, Y.M., & Tay, J.H. (2003a). Biosorption kinetics of cadmium(II) on aerobic granular sludge. *Process Biochemistry*, *38*, 997–1001.

Liu, Y., Xu, H., Yang, S.F., & Tay, J.H. (2003b). A general model for biosorption of Cd^{2+}, Cu^{2+} and Zn^{2+} by aerobic granules. *Journal of Biotechnology*, *102*, 233–239.

Liu, Y., & Tay, J.H. (2004a). State of the art of biogranulation technology for wastewater treatment. *Biotechnology Advances*, *22*, 533–563.

Liu, Y., Xu, H., Yang, S.F., & Tay, J.H. (2004b). A theoretical model for biosorption of cadmium, zinc and copper by aerobic granules based on initial conditions. *Journal of Chemical Technology and Biotechnology*, *79*, 982–986.

Liu, Y., Wang, Z.W., Liu, Y.Q., Qin, L., & Tay, J.H. (2005). A generalized model for settling velocity of aerobic granular sludge. *Biotechnology Progress*, *21*, 621–626.

Lodi, A., Solisoio, C., Converti, A., & Del Borghi, M. (1998). Cadmium, zinc, copper, silver and chromium (III) removal from wastewaters by *Sphaerotilus natans*. *Bioprocess Engineering*, *19*, 197–203.

Mohan, D., & Singh, K.P. (2002). Single- and multi-component adsorption of cadmium and zinc using activated carbon derived from bagasse – an agricultural waste. *Water Research*, *36*, 2304–2318.

Othman, M.R., & Amin, A.M. (2003). Comparative analysis on equilibrium sorption of metal ions by biosorption Tempe. *Biochemical Engineering Journal*, *16*, 361–364.

Pethkar, A.V., Kulkarni, S.K., & Paknikar, K.M. (2001). Comparative studies on metal biosorption by two strains of *Cladosporium caldosporioides*. *Bioresource Technology*, *80*, 211–215.

Puranik, P.R., Modak, J.M., & Paknikar, K.M. (1999). A comparative study of the mass transfer kinetics of metal biosorption by microbial biomass. *Hydrometallurgy*, *52*, 189–197.

Scott, J.A., & Karanjkar, A.M. (1992). Repeated cadmium biosorption by regenerated *Enterobacter aerogenes* biofilm attached to activated carbon. *Biotechnology Letters*, *14*, 737–740.

Singh, S., Rai, B.N., & Rai, L.C. (2001). Ni(II) and Cr(VI) sorption kinetics by Microcystis in single and multimetallic system. *Process Biochemistry*, *36*, 1205–1213.

Taniguchi, J., Hemmi, H., Tanahashi, K., Amamo, N., Nakayama, T., & Nishino, T. (2000). Zinc biosorption by a zinc-resistant bacterium, *Brevibacterium sp*. strain HZM-1. *Applied Microbiology and Biotechnology*, *54*, 581–588.

Valdman, E., & Leite, S.G.F. (2000). Biosorption of Cd, Zn and Cu by *Saragssum sp*. waste biomass. *Bioprocess Engineering*, *22*, 171–173.

Wang, J.L., Zhao, X.M., Ding, D.C., & Zhou, D. (2001). Biosorption of lead(II) from aqueous solution by fungal biomass of *Aspergullus niger*. *Journal of Biotechnology*, *87*, 272–277.

Xu, Y.C. (2002). Biosorption of heavy metals by *Laminaria japonica*. *Ph.D. Thesis*, The University of Texas at Arlington.

Xu, H. (2006). Equilibrium, thermodynamics and mechanisms of heavy metal biosorption by aerobic granules. *Ph.D. Thesis*, Nanyang Technological University, Singapore.

Xu, H., Tay, J.H., Foo, S.K., Yang, S.F., & Liu, Y. (2004). Removal of dissolved copper and zinc by aerobic granular sludge. *Water Science and Technology*, *50*, 155–160.

Xu, H., Liu, Y., & Tay, J.H. (2005). Effect of pH on nickel biosorption by aerobic granular sludge. *Bioresource Technology*, *97*, 359–363.

Zhou, J.L. (1999). Zn biosorption by *Rhizopus arrhizus* and other fungi. *Applied Microbiology and Biotechnology*, *51*, 686–693.

Chapter 12

Conclusions: Current State and Directions of Research

The Development of Anaerobic Granulation

The anaerobic granulation system has been known for its unique ability to convert highly objectionable wastes into useful products. With global concerns over energy shortages and greenhouse gas formation through combustion of fossil fuels, more efforts towards renewable energy supplies is clearly needed. Greater efforts are now needed for broader applications of anaerobic granulation system for relieving the environment of unwanted organic materials by converting them into methane, a renewable energy source. The anaerobic granulation process leading towards efficient methane production from wastewaters clearly fits this need.

At the moment, the most popular treatment process is the UASB reactor. However, with the recent development of EGSB and "staged multi-phase anaerobic" (SMPA) reactor systems, this may lead to a very promising new generations of anaerobic treatment system. These concepts behind the EGSB will provide a higher efficiency at higher loading rates, are applicable for extreme environmental conditions (e.g. low and high temperatures) and to inhibitory compounds. Moreover, by integrating the anaerobic process with other biological methods (sulfate reduction, microaerophilic organisms) and with physical–chemical methods, a complete treatment of the wastewater can be accomplished at very low costs, while at the same time valuable components can be recovered for reuse. Anaerobic treatment has developed into an established technology for a wide variety of industrial applications. As the waste strength tends to increase for industrial effluents, there is a need for development of anaerobic granular biomass-retaining reactors. The membrane bioreactors (MBR) with granular-based anaerobic processes may offer a solution for specific wastewater treatment which are worth exploring into.

Environmental regulations are oriented towards the sustainability of the production processes, and this leads to better recovery of resources from raw materials and by-products, energy saving, and so on. Granular sludge-based anaerobic processes have been receiving widespread recognition in their ability to offer high degree of organics removal, low sludge production, and low energy consumption along with energy production in the form of biogas. It may not be an unreasonable expectation that, in the future, the wastewater treatment technologies will experience a global shift towards usage of highly efficient granular sludge-based anaerobic processes.

Mechanisms of Aerobic Granulation

Although extensive work has been done in the area of aerobic granulation, future research needs to look into different aspects of physiology, ecology, and molecular biology of microbial granules. Further research needs to address a basic question about formation mechanisms of anaerobic and aerobic granules. It is clear that different physico-chemical mechanisms, such as hydrophobic and electrostatic cell–cell interactions, formation of polysaccharide matrix and salt bridges connecting cells, biologically specific cell-to-cell aggregation, formation of mechanically strong outer frame ("skin") and inner frame of filamentous bacteria are important in granulation but the specificity and role of these mechanisms have not been studied yet.

Physiological Diversity in Aerobic Microbial Granules

The important aspect of granulation is physiological diversity of microbial cells in the granule. For example, distribution of cells by phases of cells division cycle or mytotic cycle can produce significant information about cell activity and interactions inside the granule. It was not studied yet, the coordination between biochemical and physiological cell activities such as individual RNAs and enzymes synthesis and degradation, regulation of enzyme activity by metabolites and cofactors, regulation of catabolism and energy storage, regulation of whole-cell activity by different cell regulators during granule formation and maturation.

Distribution of Exotrophic and Endotrophic Microbial Cells in Granule

A theory explaining coordination of cell cycle events is alternation of the periods of exotrophy and endotrophy in a cell cycle (Ivanov, 2006). Cell cycle comprises the phases of exotrophy, when the external source of carbon and energy is extensively transformed into energy and carbon store (glycogen, starch, lipids), and the phases of endotrophy, when the accumulated store of energy and carbon is utilized for DNA replication and mytosis. External sources of energy and carbon are not assimilated during endotrophy periods. The alternations between the periods of exotrophy and endotrophy are performed due to the changes of intracellular concentration of cyclic AMP and are accompanied by alternation of the charge of membrane potential. Environmental factors which are unfavorable for DNA replication retain cells in phases of endotrophy. An extended period of exotrophy leads to enormous intracellular accumulation of carbon and energy sources. Exotrophic and endotrophic cells are distinguished by their biochemical and physiological properties so greatly that it would be useful to study these two different groups of cells. Therefore, distribution and percentage of exotrophic and endotrophic cells in the granule is an important information on its state and structure. For example, duration of exotrophy (Δt_{ex}) of yeasts is linearly related to the duration of cell cycle (T): $\Delta t_{\text{ex}} = 0.5T - 1.0$. Using this equation, the specific growth rate of the selected species in the granules (μ) can be determined from the microscopic view, taking into account that $T = \ln 2/\mu$. Exotrophic and endotrophic cells can be distinguished after adding a small quantity of co-oxidizing substrate, which is transformed into toxic products of oxidation. For example, allyl or amyl alcohol can be added to cells, which utilize ethanol. As a result, cells will produce allyl or amyl aldehyde, which cannot be further oxidized and, therefore, will kill cells. Exotrophic cells die after this incubation but endotrophic cells remain alive because they do not consume and oxidize external sources of carbon and energy. The share of exotrophic cells increases during starvation and other unfavorable conditions because the phase of DNA synthesis cannot be started until sufficient intracellular quantities of carbon and energy sources are accumulated. However, distribution and percentage of exotrophic and endotrophic microbial cells in the granule, was not studied yet.

Microbial Diversity of Aerobic Granules

Experimental data demonstrated diverse microbial community in the granules. The analysis of microbial community, residing in the aerobically grown granule, can provide information on the microorganisms responsible for granule formation, maintenance, and activity. This knowledge can be used for different purposes:

1. for better control of aerobic granulation;
2. for selection of safe and effective microbial inoculum for fast granulation;
3. for selection of safe and affective microbial inocula enhancing activities of microbial granules.

It would be important to study the feasibility of bioaugmentation of microbial granules by pure cultures and recombinant species of microorganisms to tailor microbial granules for treating specific types of wastewaters. There may be complex positive interactions and horizontal gene transfers in the ecosystem of microbial granule due to close cell arrangements.

Stability of Microbial Granules

Compared to anaerobic granules, aerobic granules have relatively low stability. It would be desirable to develop a practical strategy for improving the stability of aerobic granules by manipulating operational conditions or through selecting for slow-growth bacteria. The mechanisms and rates of cell attachment–detachment in the granule are not known clearly. The life cycle of the granules was not studied and it is not clear whether new granules are faster produced from the existing granules, from cells, or particles of disrupted granules.

Formation of Aerobic Microbial Granules in Continuous Systems

Aerobic granulation has been observed only in the SBRs. However, SBRs cannot replace existing aerobic tanks.Therefore, the feasibility of aerobic granulation in existing continuous systems used in wastewater treatment should be investigated. It is clear that at least two aspects must be taken into account:

1. formation of granules from cells and flocs due to shearing force and physical or physico-chemical interactions combining cells together;
2. continuous selection/recycling of cells, which are forming granules due to cell aggregation.

Microbial Seeds

Important element for fast formation of the granules and granulation in continuous systems is one time or time-to-time addition of microbial seeds. From the practice of conventional activated sludge system it is considered by the majority of the researchers that addition of disrupted, previously formed granules can be a method for facilitation of granulation. However, major obstacle for this conventional technique may be that granules are formed due to microbial cell aggregation. This can increase a risk of the presence of pathogenic microorganisms in the granule in comparison with conventional activated sludge because ability of some strain for hydrophobic or biospecific aggregation could correlate with its pathogenicity. That is why, use of selected and safe microbial strain for granulation could be the best way for application of microbial granulation in the wastewater treatment. This granule-forming strain might be complemented with some additional microbial strains performing specific functions in the granule, like biodegradation of specific compounds or removal of nutrients from the wastewater.

Practical Application of Aerobic Microbial Granules

The biodegradation activity of microbial cells in granules is smaller than in microbial flocs of activated sludge and significantly smaller than that of suspended cells. It was shown in many experiments and generally it just follows the ratio of surface of mass transfer from medium into particle to volume of this particle. Other disadvantages of the treatment of wastewater with microbial granules can be considered their potential instability, long-time formation, and risk of accumulation of pathogens. Therefore, application of aerobic microbial granules can be useful for all cases, but only for case-specific wastewater treatments, for example:

1. Land is a premium, so absence of settling stage in the wastewater treatment using microbial granules can give significant economic advantages in comparison with conventional activated sludge process;
2. Electrical energy is cheap, so intensive aeration in granulation process cannot be an obstacle in application of microbial granules;
3. There are substances in wastewater that are toxic for microbial cells; due to the presence of protective outer layer the granules are more resistant to toxicants than microbial flocs or suspended cells;
4. Granules are effective in the treatment of ammonia-containing wastewater and simultaneous removal of organic matter and nitrogen from wastewater due to the retention of nitrifiers in the granules.

Currently, there are few known cases of recently started pilot and industrial-scale applications of aerobic granules in wastewater treatment and it is impossible to analyze efficiency of practical applications from this limited experience. However, there is no doubt that new applications of granulation technology will demonstrate more cases, when the aerobic microbial granules will be more effective than commonly used microbial activated sludge.

Color Plate Section

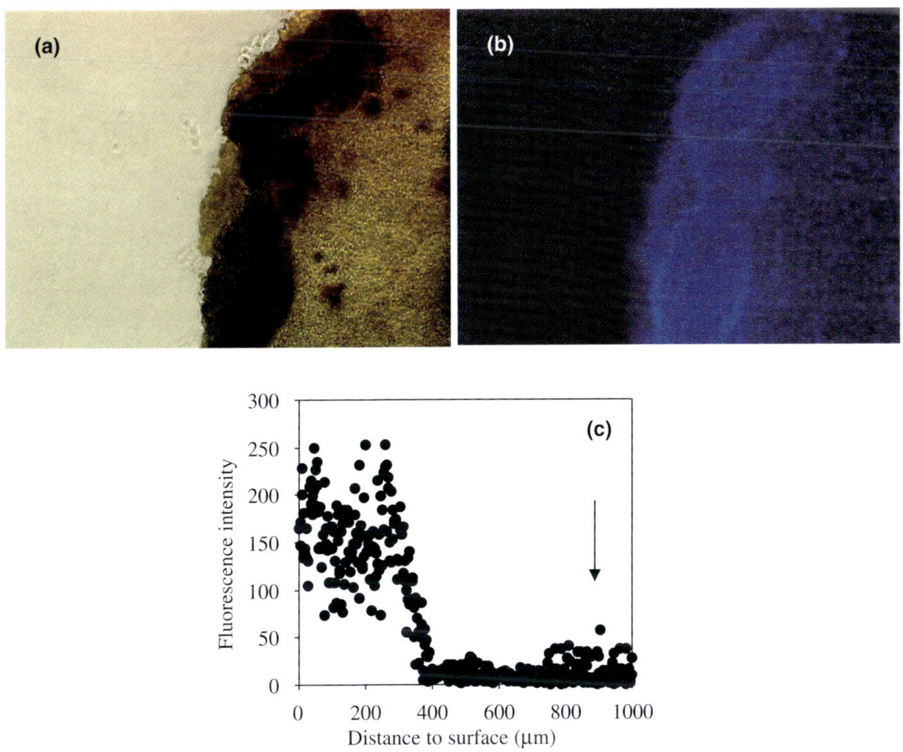

Plate 4.4. Cross section view of aerobic granules; (a) fresh granule; (b) granule stained by calcofluor white. Bar: 100 μm; (c) profile of the dye fluorescence intensity distribution along the granule radius from the surface to the center (arrow) (Wang et al., 2005b).

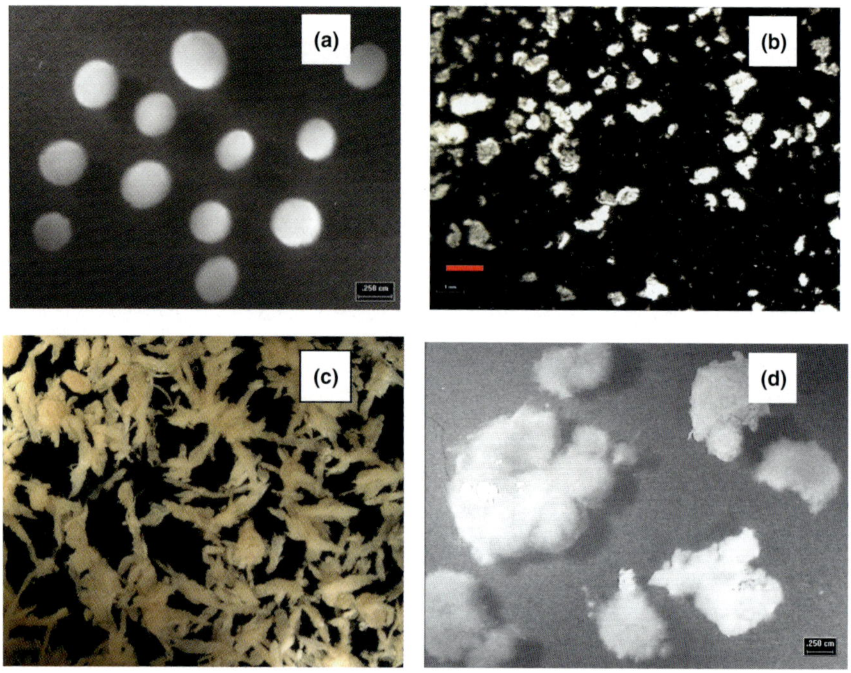

Plate 6.1. Shape of the aerobically grown microbial granules (a) spherical and ellipsoid granules; (b) granules of irregular shape; (c) super-elongated granules produced at high upflow air velocity (photo from Dr. Liu Yongqiang); (d) granules produced by filamentous microorganisms (fungi, actinomycetes, filamentous bacteria).

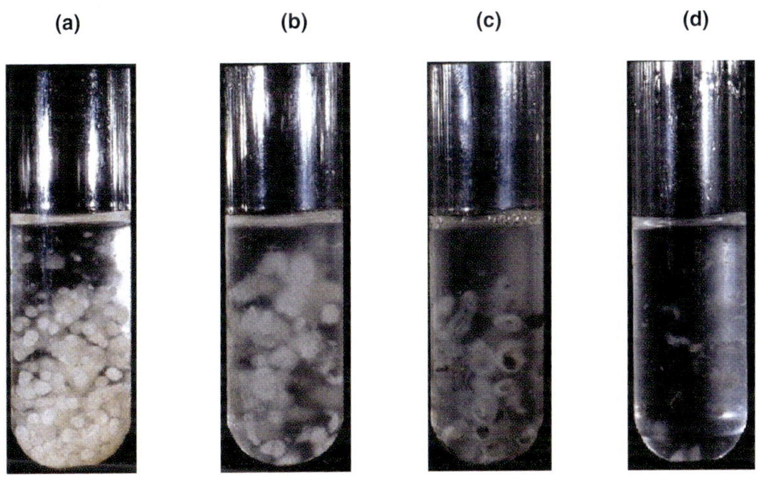

Plate 7.2. Four stages of aerobic granules development: (a) young granules; (b) mature granules; (c) old granules with black cores; and (d) disintegrated granules.

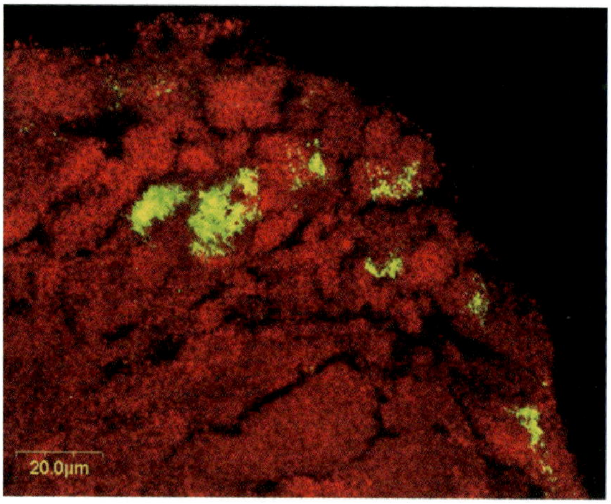

Plate 9.3. FISH–CLSM image of outer section of the granule. Red area represents cells hybridized with an eubacterial probe and green area represents cells hybridized with a probe specific for strain PG-01 (Jiang et al., 2004b).

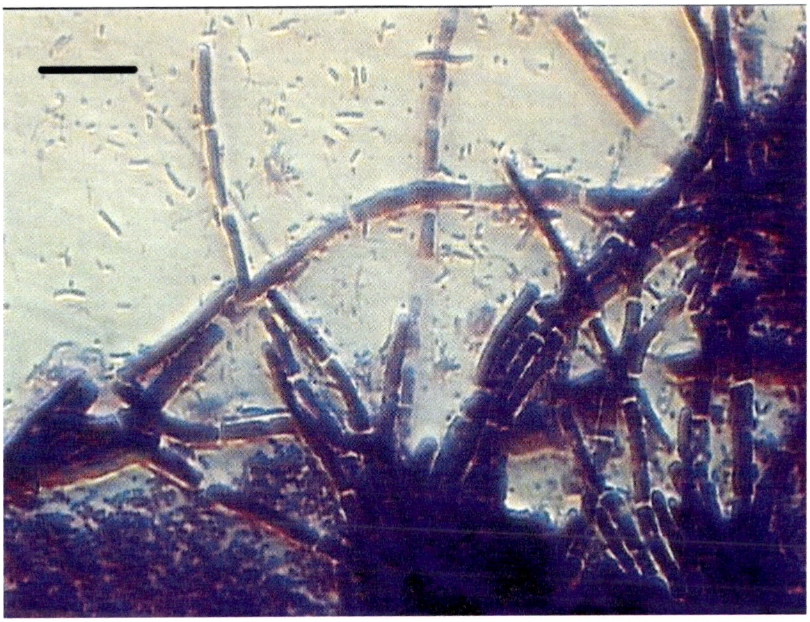

Plate 10.12. Light microscopy image of sheath bacteria on surface of phenol-degrading granule. Scale bar is 10 μm long.

Plate 10.13. DGGE profiles of R1 and R2 using partial bacterial 16S rRNA gene fragments. Lanes: 1, migration standards; 2, R1 biomass on day 0; 3, R1 biomass on day 3; 4, R2 biomass on day 0; 5, R2 biomass on day 3; 6, R2 biomass on day 15; 7, R2 biomass on day 30; 8, migration standards.

Index